Structural Analysis I
Homework and Exams

Peter I. Kattan

Petra Books
www.PetraBooks.com

Peter I. Kattan, PhD

Correspondence about this book may be sent to the author at one of the following two email addresses:

pkattan@petrabooks.com

info@petrabooks.com

Structural Analysis I. Homework and Exams
ISBN: 979-8-3492-8568-4

Preface

These are the homework sets and exam sets for the course Structural Analysis I that was taught by Dr. Peter I. Kattan. These homework and exam sets are a follow-up volume to the successful volume Structural Analysis I Lecture Notes that is available online and at fine bookstores.

This volume includes three homework sets and nine exam sets. Solutions in detail are provided for some of these sets. In particular, detailed handwritten solutions are provided for four homework sets and for five exam sets. Note that these sets as well as the associated Lecture Notes are for the first course in structural analysis that covers determinate structures. Another set of notes as well as homework and exams are available for the second course on indeterminate structures.

The author makes these notes and sets available as free downloads as PDF files on the internet. However, the printed paperback copies are provided at a greatly reduced price at fine bookstores online and everywhere. The total size of this volume is about 80 page. Not all the solutions are provided but enough is given to keep the student busy for hours.

April 2025 Peter I. Kattan

Structural Analysis I
Homework and Exams

Contents

Structural Analysis I

Homework and Exams

APPLIED SCIENCE UNIVERSITY
FACULTY OF ENGINEERING

Structural Analysis (1)

Homework 8

1- The beam shown in the figure is subjected to the motion of a series of seven wheel loads of a multiaxle truck. Determine the critical position of the loads for maximum R_A, then compute the maximum value of R_A.

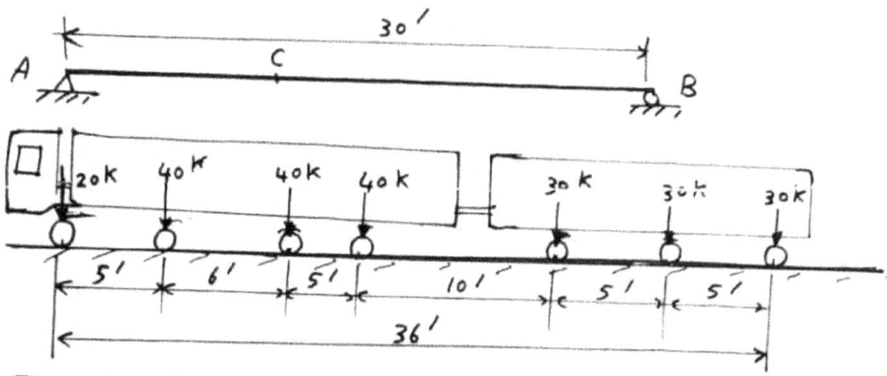

2- The seven axle semitrailer is to pass over the same 30-ft simple beam shown above. Determine the critical position of the wheels for maximum shear at C, then compute the maximum positive value of V_C. Point C is located 10 ft from the left end A.

3- A series of nine fixed-spaced wheel loads are assumed to move from right to left on the simply-supported beam shown in the figure. Determine the position of the wheels that will give maximum moment at C, then compute the maximum value of M_C.

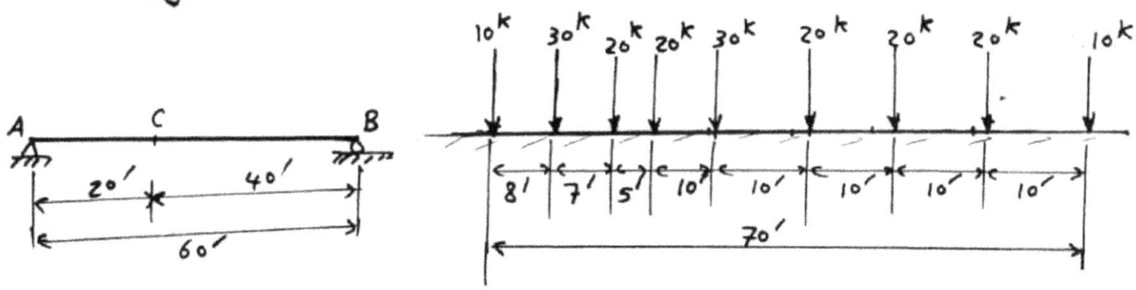

4- (a) Determine the maximum moment at midspan of the 20-ft beam due to the system of wheel loads shown in the figure.
(b) Determine the absolute maximum moment in the 20-ft simply-supported beam due to the system of wheel loads shown in the figure.

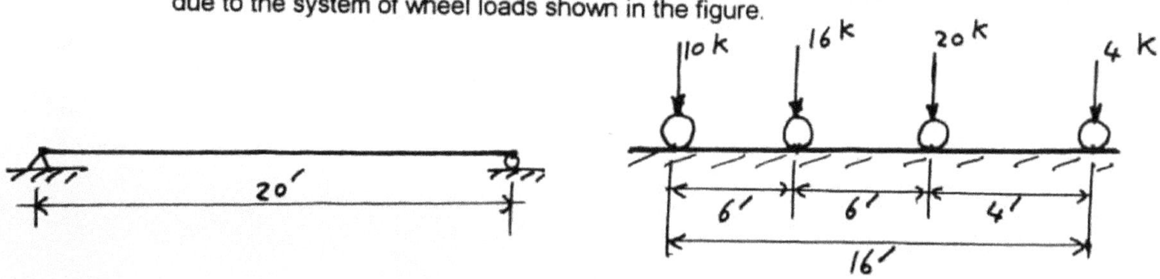

APPLIED SCIENCE UNIVERSITY
FACULTY OF ENGINEERING

Structural Analysis (1)

Homework 9

1- Problem 9.2 page 215 in the textbook (Alexander Chajes).

2- Problem 9.4 page 216 in the textbook (Alexander Chajes).

3- Draw the normal force, shear force, and bending moment diagrams for the semicircular arch shown in the figure.

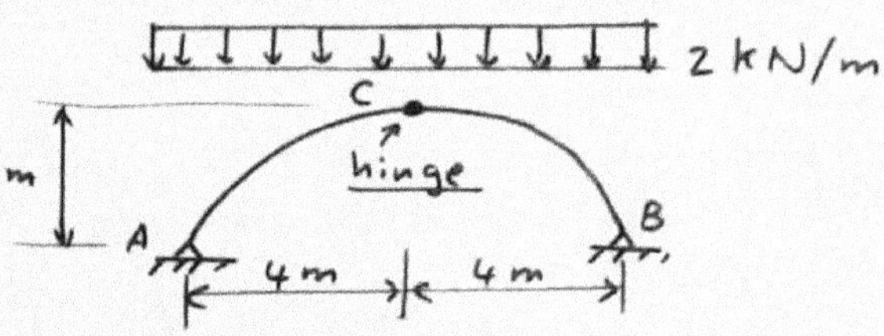

4- Find the normal force, shear force, and bending moment at section 1-1 for the three-hinged parabolic arch shown in the figure. The arch equation is $y = x - 0.025x^2$.

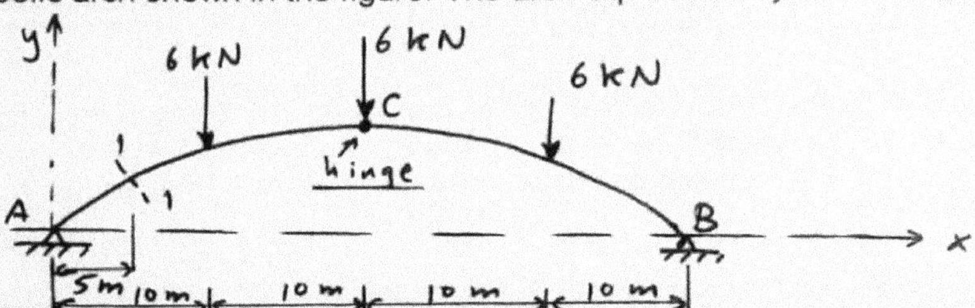

5- Calculate the horizontal and vertical deflections of the free end (point C) of the structure shown in the figure. The structure is composed of a vertical member AB and a semicircular member BC. Use E = 210,000 MPa and I = 369.70 x10^6 mm^4.

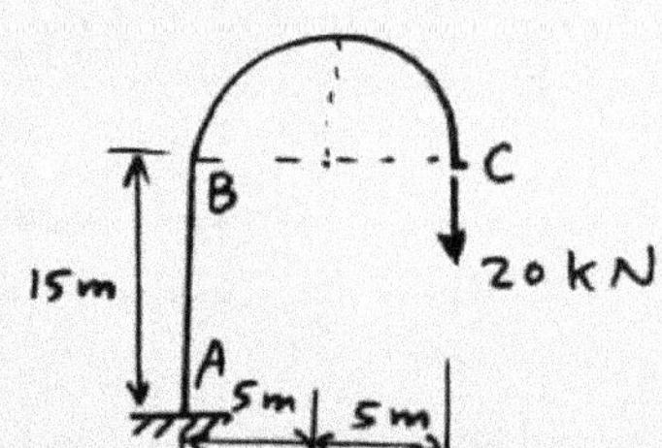

6- Draw influence lines for the vertical and horizontal reactions at A, and the internal forces (normal force, shear, and moment) at section 1-1 for the semicircular arch shown in the figure.

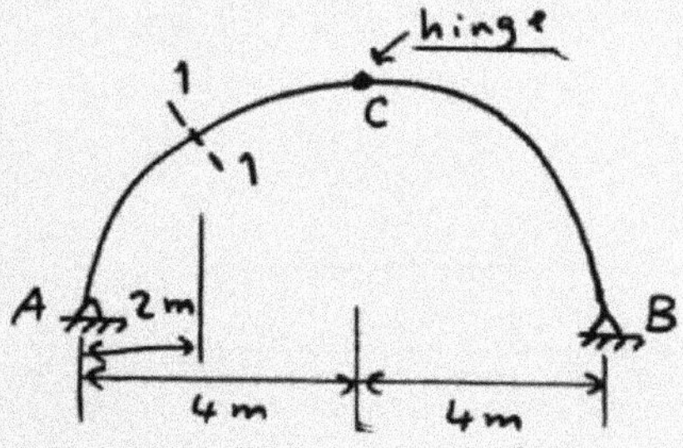

APPLIED SCIENCE UNIVERSITY
FACULTY OF ENGINEERING

Structural Analysis (1)

Homework 10

1- Determine (approximately) the support reactions at A and B due to the uniform vertical load acting on the frame shown in the figure.

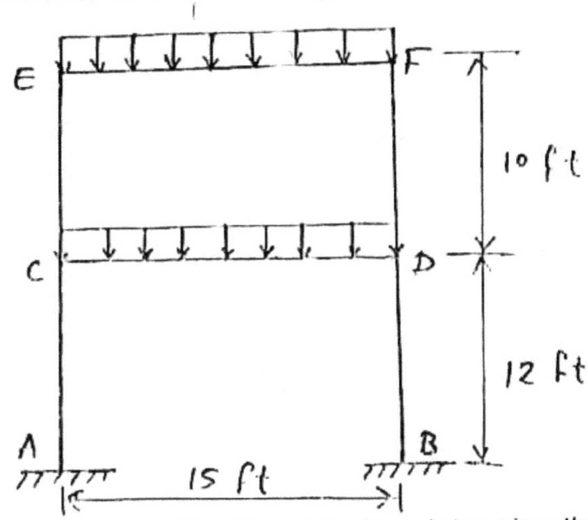

2- Using the portal method of approximate analysis, determine the reactions at the base of the columns of the frame shown in the figure. Show the results on a free body diagram of the frame.

3- Using the cantilever method of approximate analysis, determine the reactions at the base of the columns of the frame shown in the figure. Show the results on a free-body diagram of the frame.

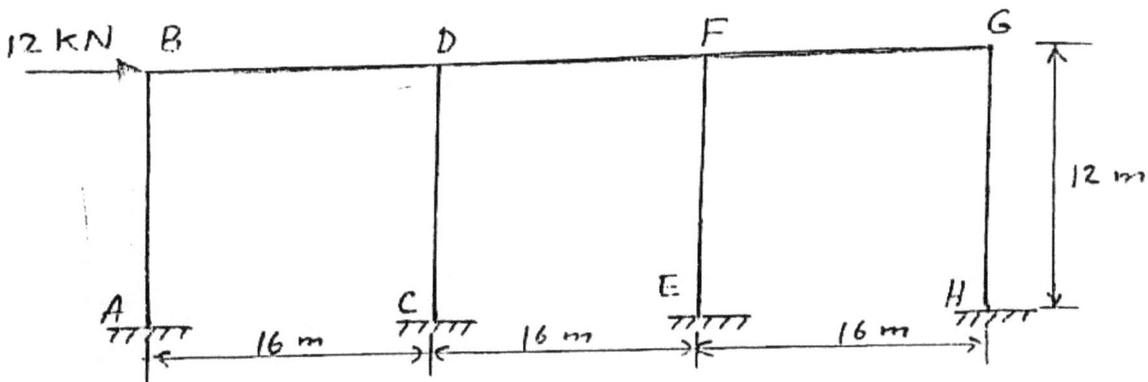

Compare the results of problems 2 and 3 with the computer results given below:

$A_x = -2.63$, $A_y = -1.43$, $M_A = 19.1$, $C_x = -3.37$, $C_y = 0.41$, $M_C = 22.0$, $E_x = -3.37$, $E_y = -0.41$, $M_E = 22.0$, $H_x = -2.63$, $H_y = 1.43$, $M_H = 19.1$

Applied Science University
Faculty of Engineering

Dr. Peter I. Kattan
Time: 1:20 hours

Structural Analysis I
Exam 1

1- (15 points) Calculate the reactions for the structure shown in the figure.

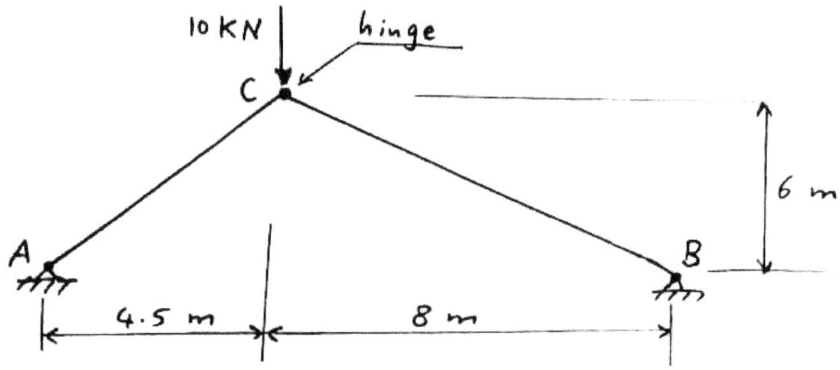

2- (20 points) Discuss the stability and determinacy of the following structures.

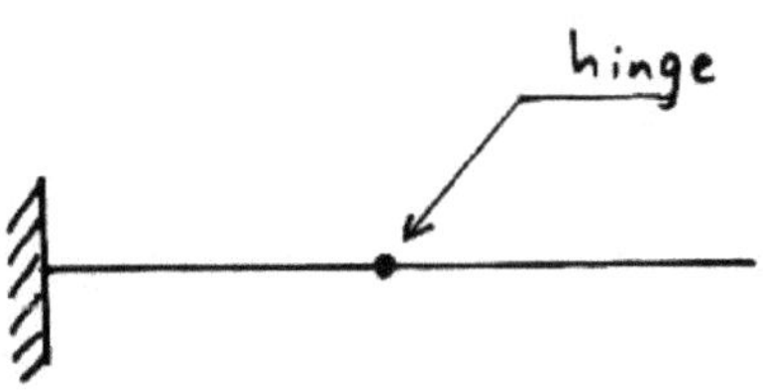

(a)

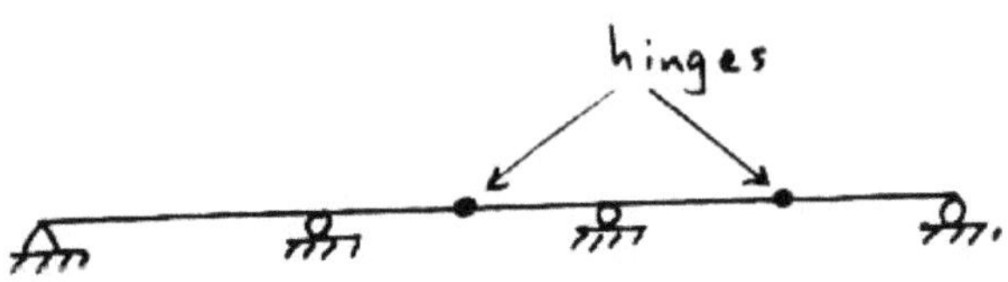

(b)

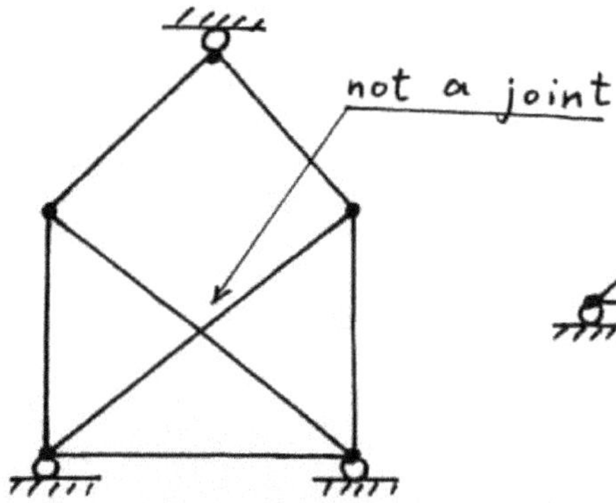

(c)

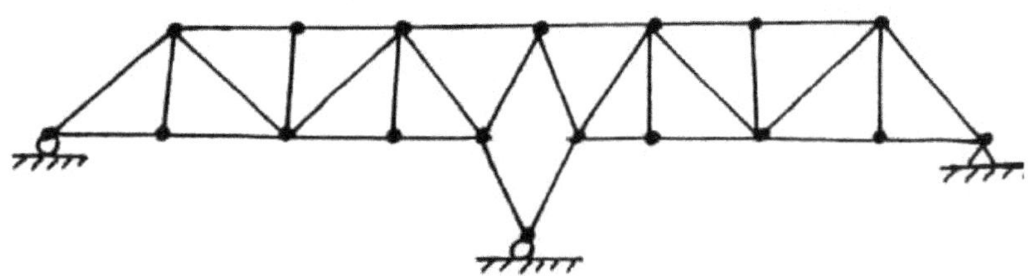

(d)

2

3- (30 points) Using the method of joints and sections combined, determine the forces in members OB, BE, BC, CE, and DE of the K-truss shown in the figure Indicate whether the members are in tension or compression.

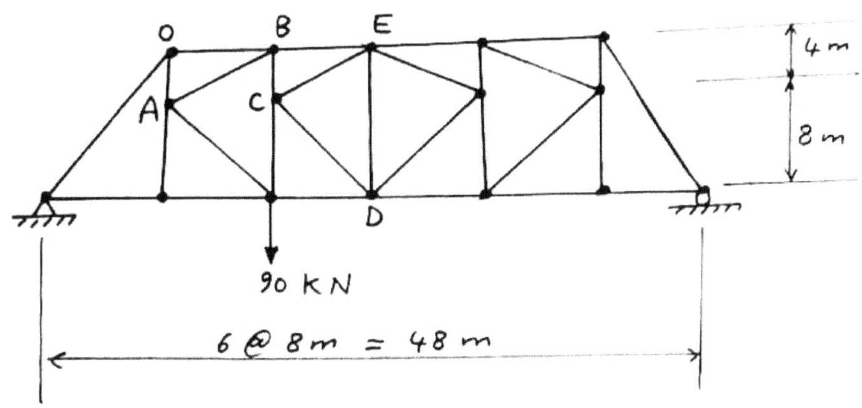

4- (35 points) Draw the shear force and bending moment diagrams for the frame shown in the figure.

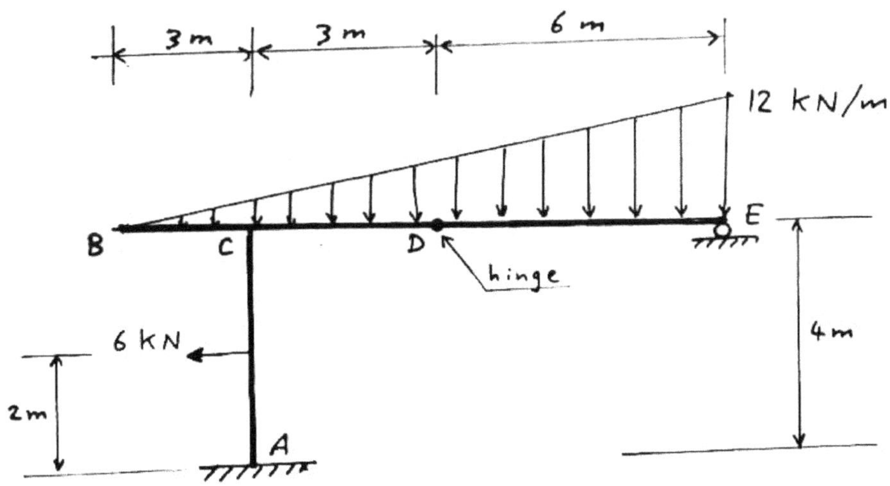

Bonus : (+5 points) Draw the shear force and bending moment diagrams for the structure of problem (1).

Note : This is a trick question.

Applied Science University
Faculty of Engineering

Dr. Peter I. Kattan
Time: 1:15 hours

Structural Analysis (1)
Exam 1

1- (40 points) Calculate the forces in members DG, EG, EI, AB, and MN of the K-truss shown in the figure. Indicate whether each member is in tension or compression.

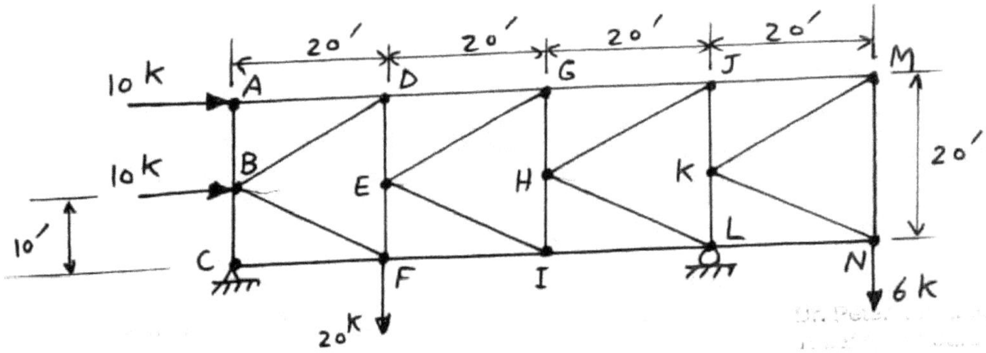

2- (40 points) Draw the shear force and bending moment diagrams for the compound beam shown in the figure.

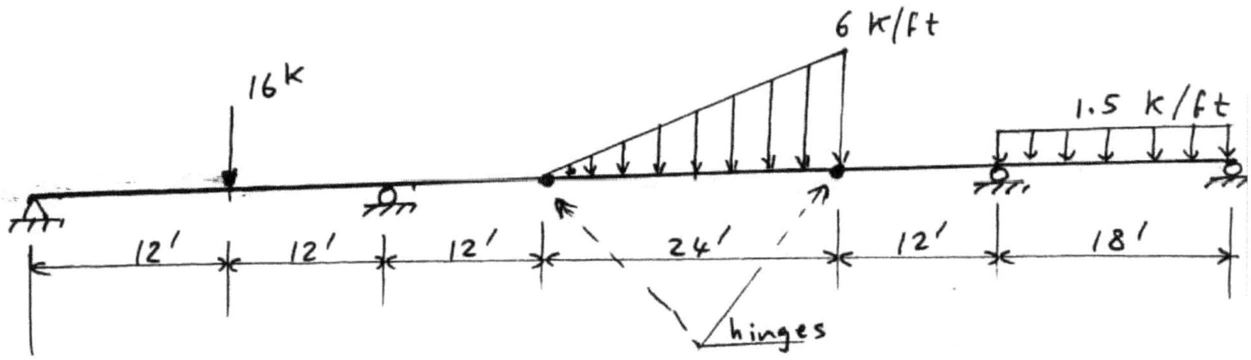

3- (20 points) Determine the slope (rotation) at A and B, and the deflection at A in the cantilever beam shown in the figure. Take $E = 1.4 \times 10^6$ psi, $I = 1200$ in^4.

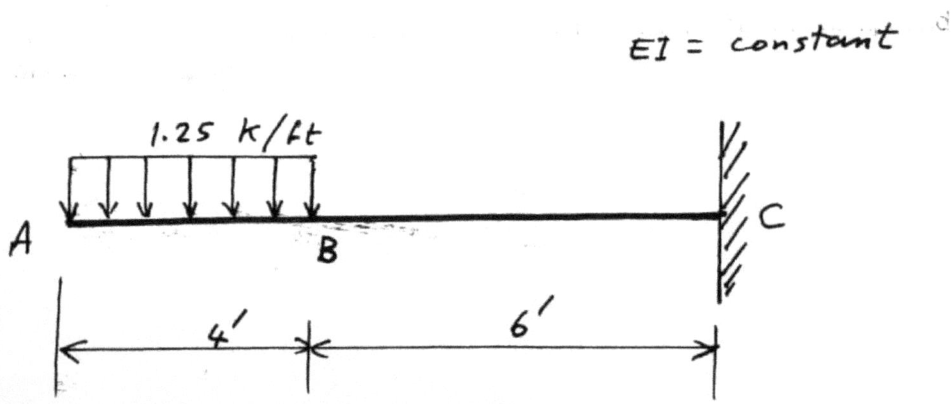

Applied Science University
Faculty of Engineering

Dr. Peter I. Kattan
Time: 1:15 hours

Structural Analysis (1)
Exam 2

1- (40%) Use the conjugate beam method to determine the deflection and rotation at point B of the compound beam shown in the figure. Take E = 20,000 kN/cm^2 and I = 15,000 cm^4.

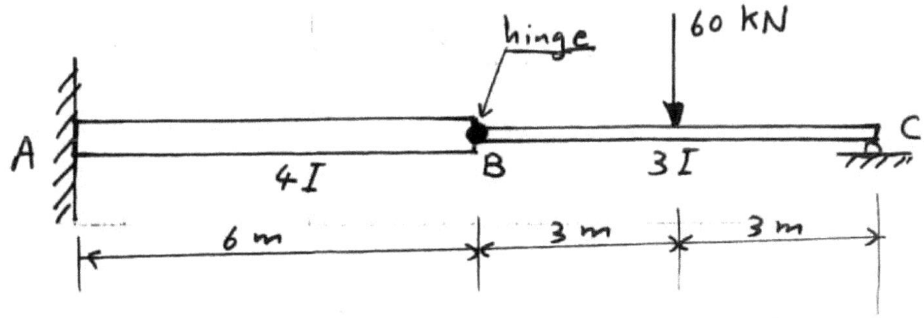

2- (30%) Determine the horizontal deflection at joint B of the truss shown in the figure. Take E = 20,000 kN/cm^2 and A = 20 cm^2.

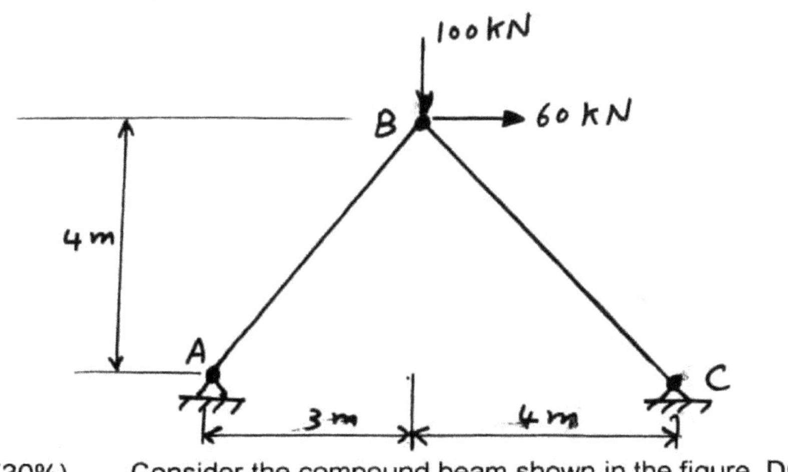

3- (30%) Consider the compound beam shown in the figure. Draw the influence lines for:

(a) the reaction at A.
(b) the shear and moment at E.

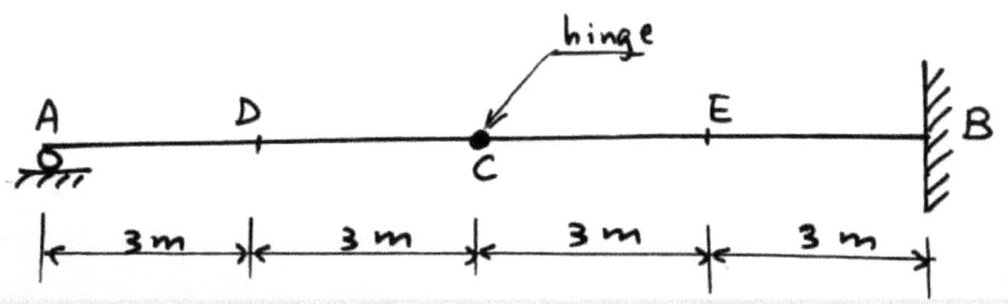

Applied Science University
Faculty of Engineering

Dr. Peter I. Kattan
Time: 1:15 hours

Structural Analysis (1)
Exam 2

1- (40%) Use the conjugate beam method to determine the deflection and rotation at point B of the compound beam shown in the figure. Take $E = 20,000\ kN/cm^2$ and $I = 15,000\ cm^4$.

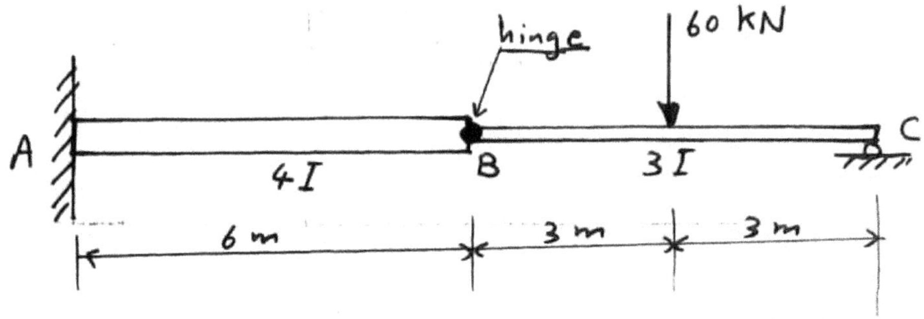

2- (30%) Determine the horizontal deflection at joint B of the truss shown in the figure. Take $E = 20,000\ kN/cm^2$ and $A = 20\ cm^2$.

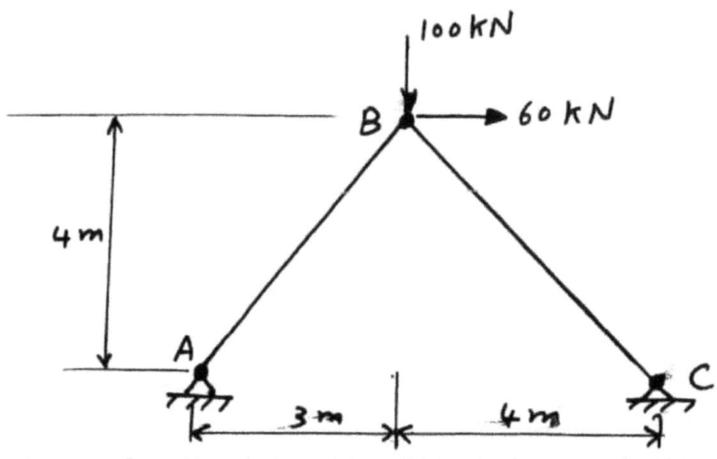

3- (30%) Consider the compound beam shown in the figure. Draw the influence lines for:

(a) the reaction at A.
(b) the shear and moment at E.

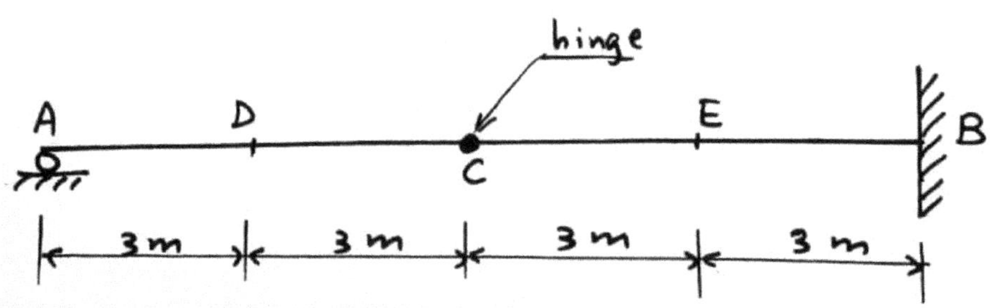

Applied Science University
Faculty of Engineering

Dr. Peter I. Kattan
Time:

Structural Analysis I
Exam 2

1- (30 points) Using the moment-area theorems, calculate the slope and deflection at the free end (point C) of the cantilever beam shown in the figure. The modulus of elasticity is constant throughout the beam. Specify the directions for the slope and deflection.
$E = 29000$ ksi, $I_1 = 1000$ in^4, $I_2 = 2000$ in^4

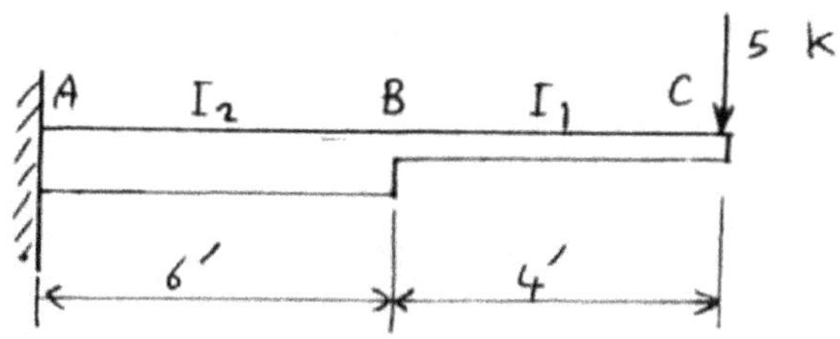

2- (35 points) Using the conjugate-beam method, calculate the slope and deflection at point C of the overhanging beam shown in the figure. Both the moment of inertia and modulus of elasticity are constant throughout the beam. Specify the directions for the slope and deflection. $E = 29000$ ksi, $I = 1400$ in^4

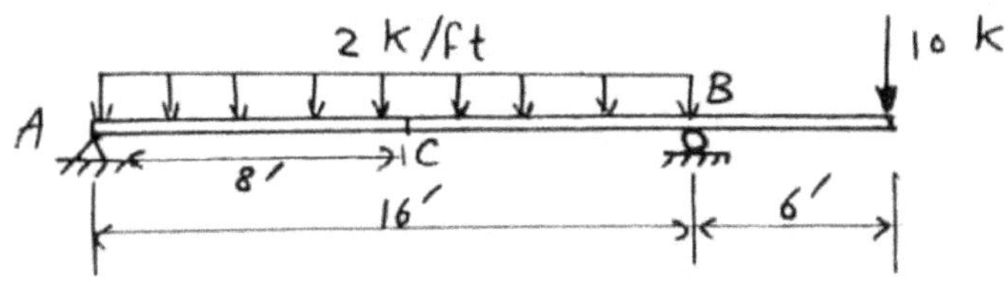

3- (35 points) Using the method of virtual work (unit-load method), calculate the horizontal displacement of the roller support at D of the truss shown in the figure. Specify the direction of motion. The modulus of elasticity ($E = 29000$ ksi) is constant for all the members. The cross-sectional area of each diagonal member is 1.5 in^2 while the cross-sectional area of each horizontal member is 0.70 in^2
(Hint: The truss is symmetric).

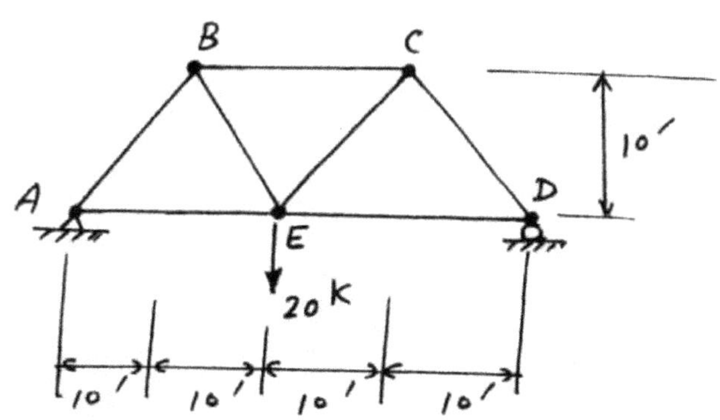

Applied Science University
Faculty of Engineering

Dr. Peter I. Kattan
Time: 1 hour

Structural Analysis (1)
Exam 3

1- (20%) Draw the influence line for the force in member AB of the truss shown in the figure.

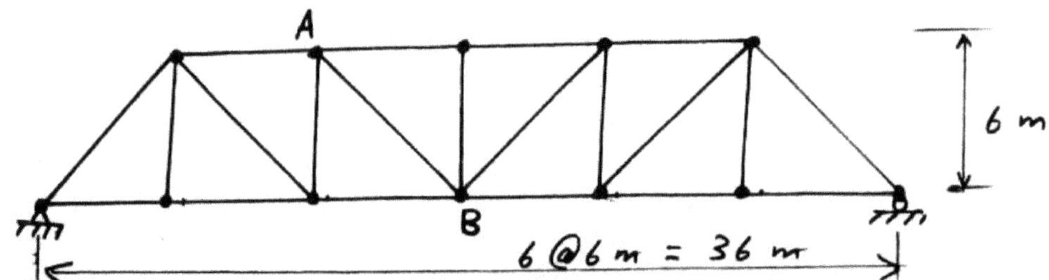

2- (40%) The simply-supported beam shown in the figure is subjected to a set of two concentrated moving loads. Determine the position of the loads and the resulting value of the absolute maximum moment in the beam.

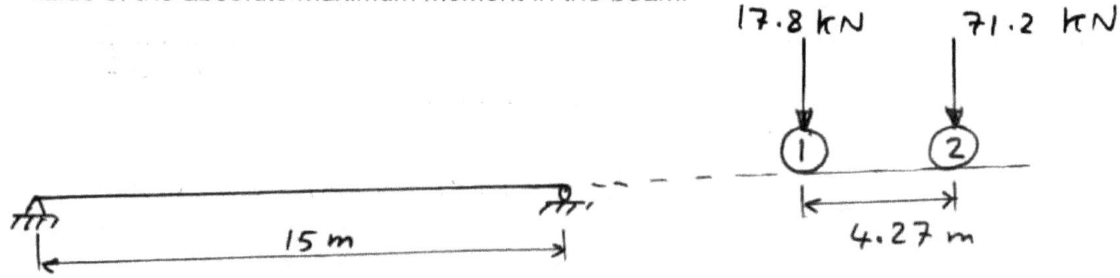

3- (40%) The slopes of the two segments of the three-hinged arch are described by the functions shown in the figure. The two supports are not on the same level. Determine:

(a) The horizontal and vertical reactions at the supports.
(b) The moment in the arch at point E.

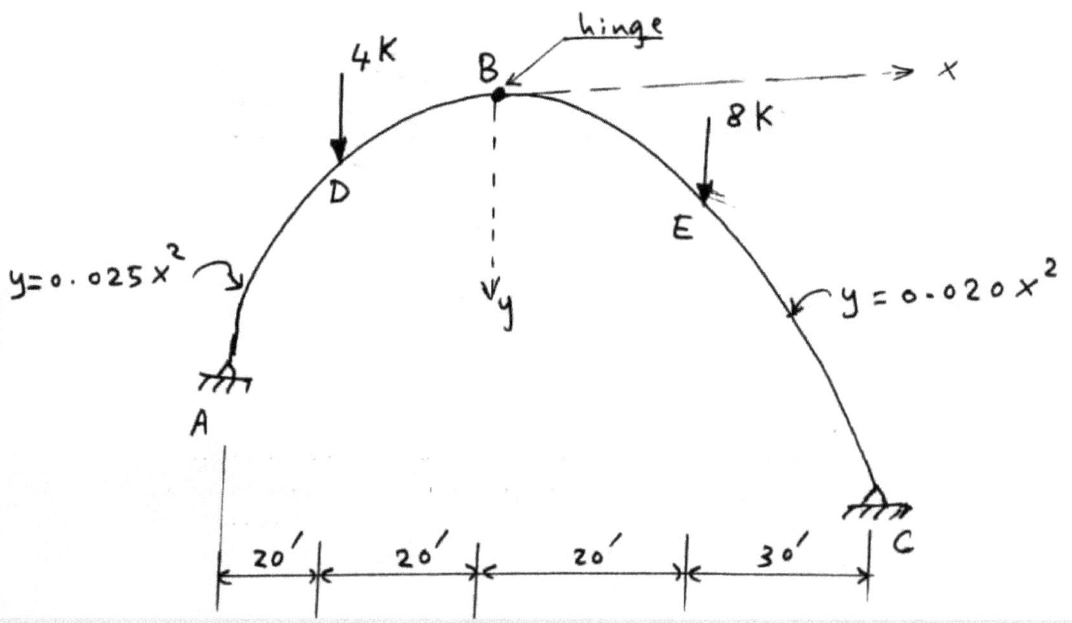

Applied Science University
Faculty of Engineering

Dr. Peter I. Kattan

Time: 1:30 hours

Structural Analysis I
Exam 3

1- (40 points) Consider the compound beam shown in the figure. Draw the influence lines for:
20 (a) the reactions at support A.
20 (b) the shear force and bending moment at point B.

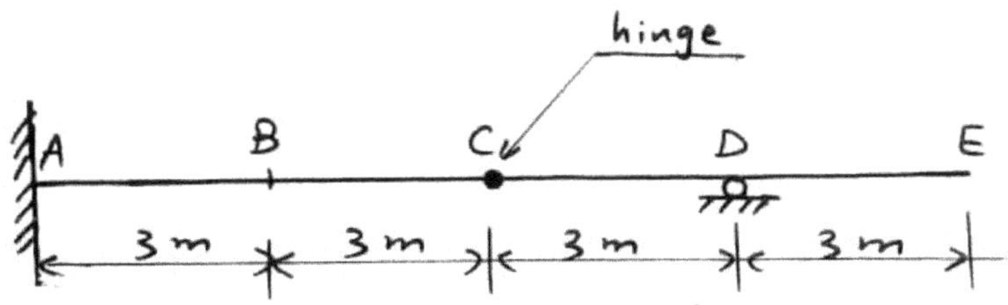

2- (30 points) Consider the simply-supported beam and the wheel loads of the car shown in the figure. Calculate:
15 (a) the maximum moment at midspan of the beam.
15 (b) the absolute maximum moment and the exact position of the car on the beam.

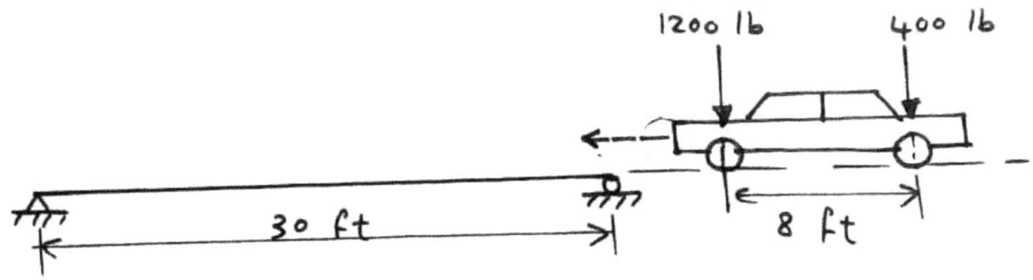

3- (30 points) For the symmetric three-hinged arch loaded as shown in the figure determine:
15 (a) the horizontal and vertical reactions at the supports A and C.
15 (b) the normal force, shear force, and bending moment at point D.

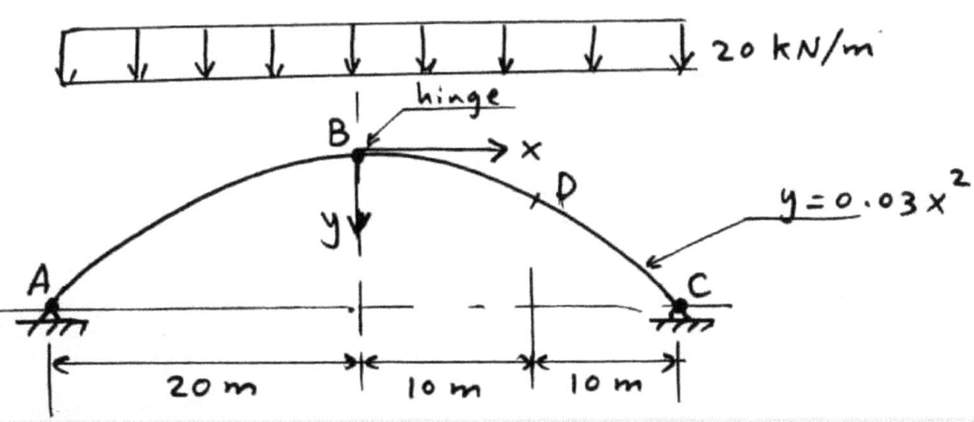

Applied Science University
Faculty of Engineering

Dr. Peter I. Kattan
Time: 1 hour

Structural Analysis I
Exam 3-a

1- (50 points) Consider the compound beam shown in the figure. Draw the influence lines for:

10 (a) the reaction at support A.
2° (b) the shear force and bending moment at point D.
2o(c) the shear force and bending moment at point E.

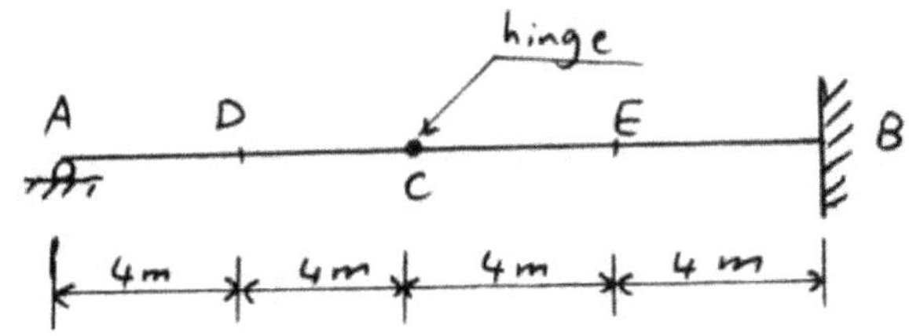

2- (30 points) For the simply-supported beam shown in the figure, find the absolute maximum bending moment due to a truck, which is represented by the series of concentrated loads shown in the figure, and which moves from the right to the left.

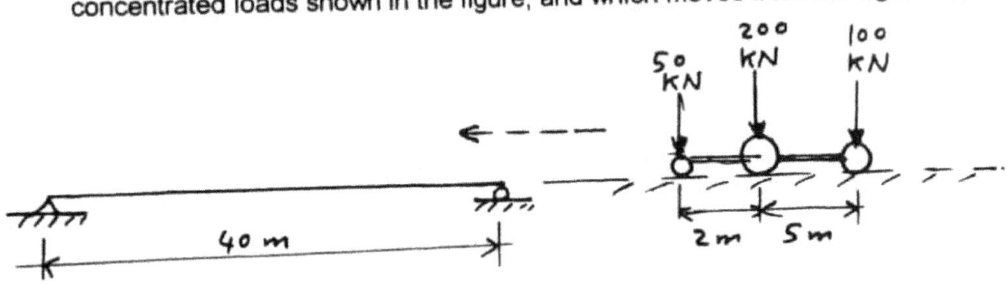

3- (20 points) Calculate the reactions for the three-hinged arch shown in the figure.

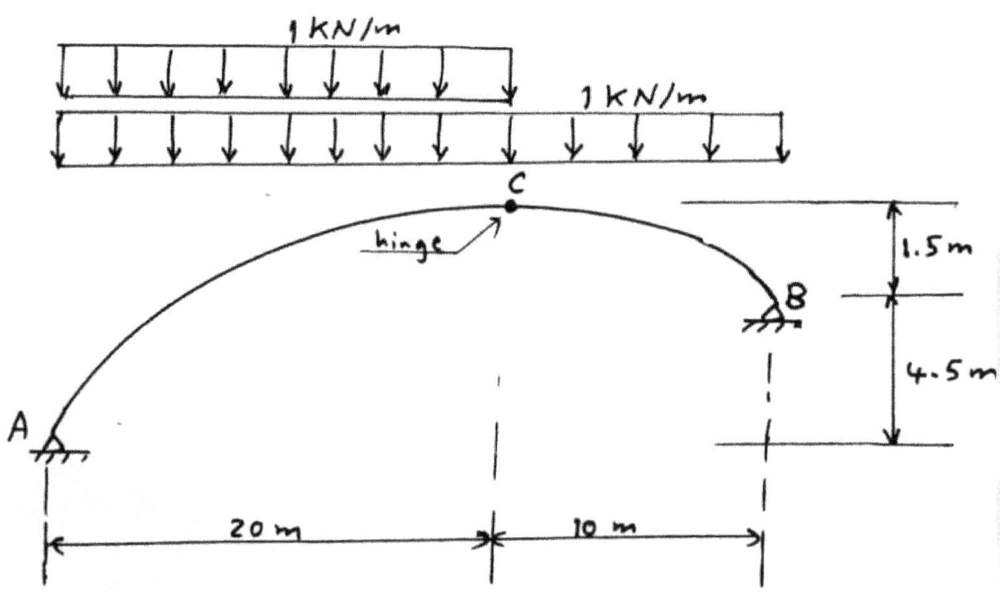

Applied Science University
Faculty of Engineering
23-1-1997

Prof. Maffuz El-Rayyes
& Dr. Peter I. Kattan
Time: 2 hours

Structural Analysis (1)
Final Exam

Answer all questions:

1- a- (12%) Determine the forces in members BC, AD, FG, and DE of the truss shown in the figure below. Indicate whether each member is in tension or compression.

b- (8%) Draw the influence line for the force in member BC.

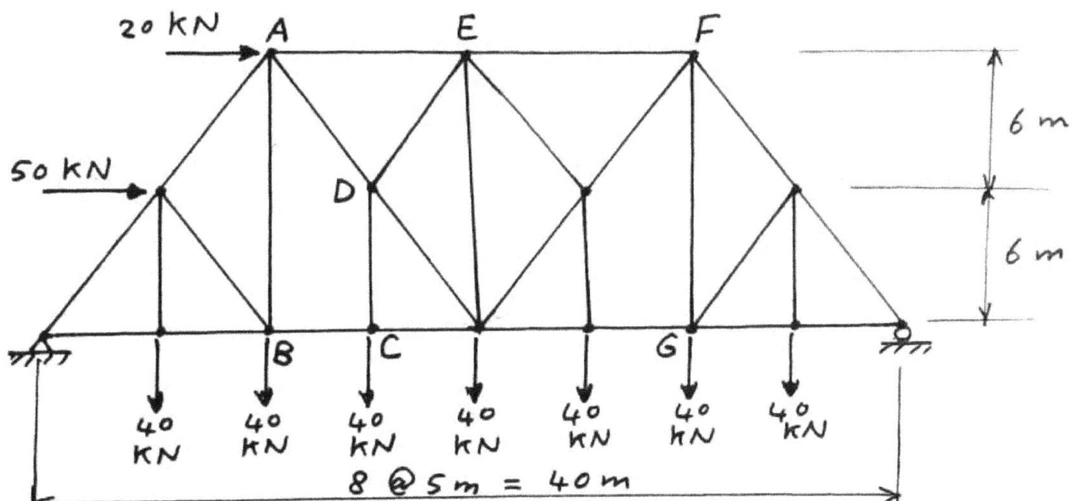

2- (25%) Determine the deflections at A and C as well as the slope (rotation) at A for the overhanging beam shown in the figure below. Take EI = 15000 kN.m^2 (indicate the directions for all results).

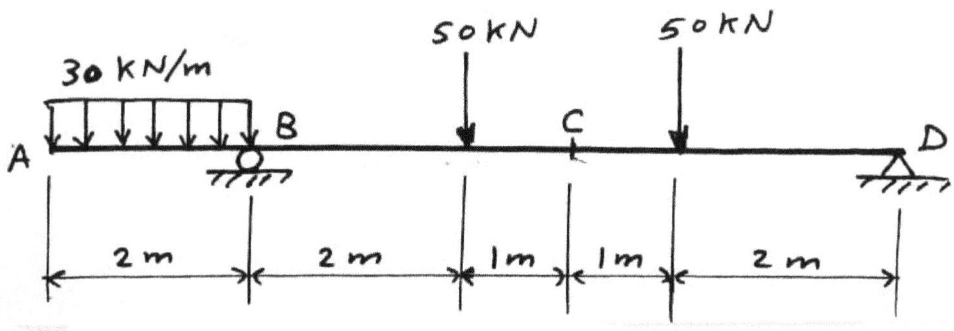

3- Consider the three-hinged arch shown in the figure below whose centerline is giver by the equation $y = 0.04(32-x)x$.

a- (12%) Determine the normal (axial) force, shear force, and bending moment at section k.

b- (13%) Draw the influence line for the horizontal reaction H, then determine the maximum value of H due to the passage of the system of moving concentrated loads shown in the figure below.

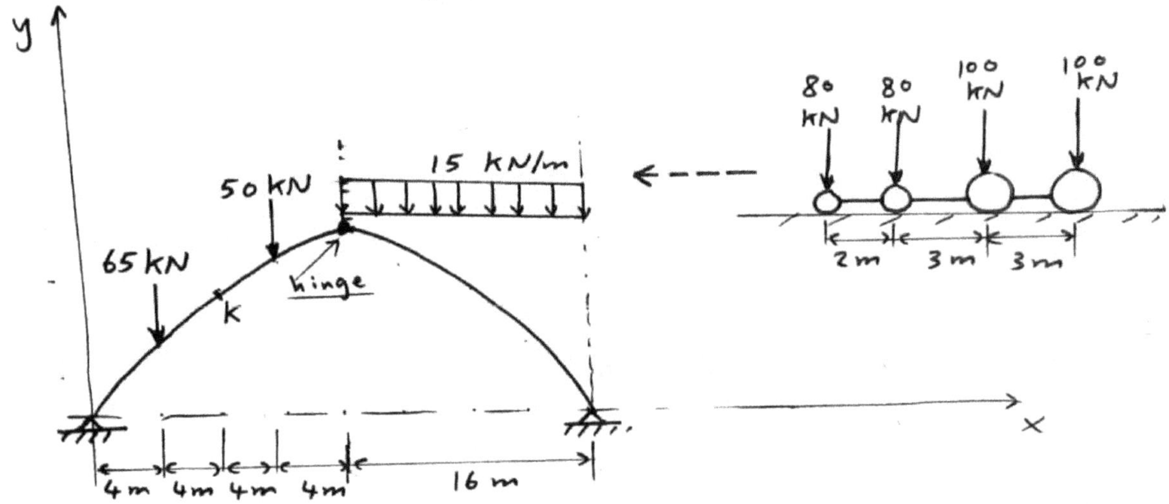

4- (30%) Draw the shear force and bending moment diagrams for **<u>only one</u>** of the two redundant structures shown in the figure below.

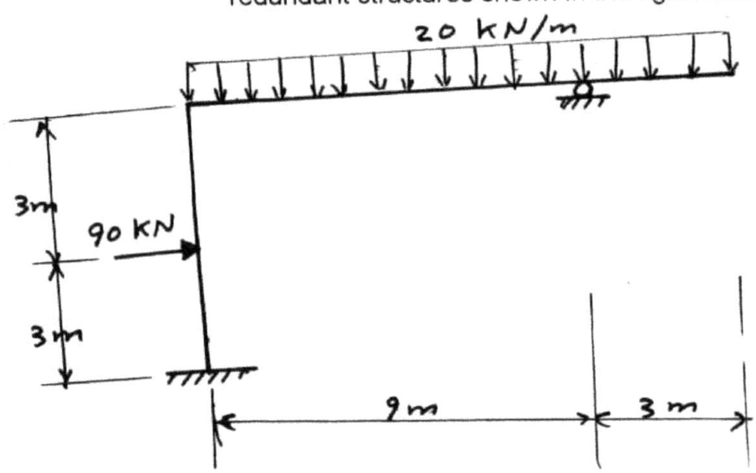

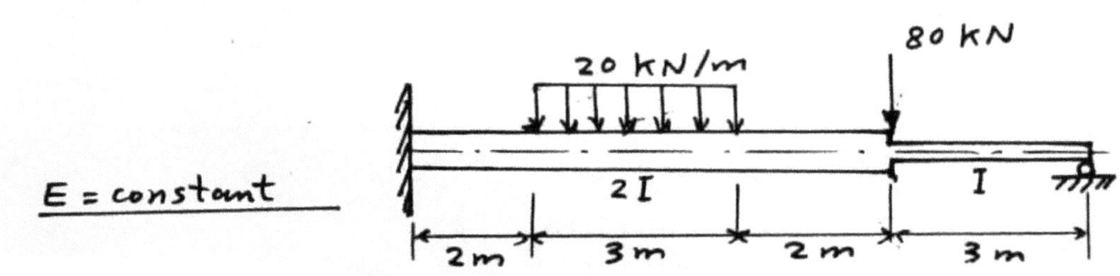

Applied Science University
Faculty of Engineering
21-6-1997

Dr. Peter I. Kattan
Time: 2 hours

Structural Analysis (1)
Final Exam

Answer all questions:

1- (15%) Calculate the reactions and the force in member AB of the truss shown in the figure.

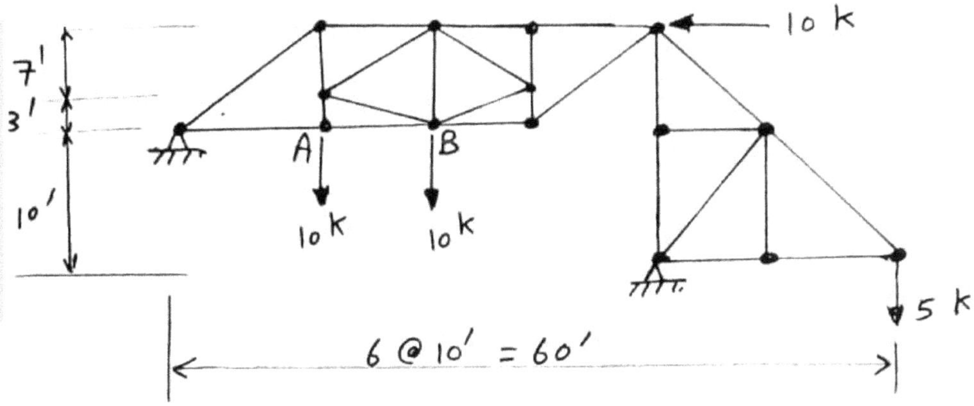

2- (15%) Calculate the magnitude of the upward force P such that the deflection at midspan is zero. Take EI = constant.

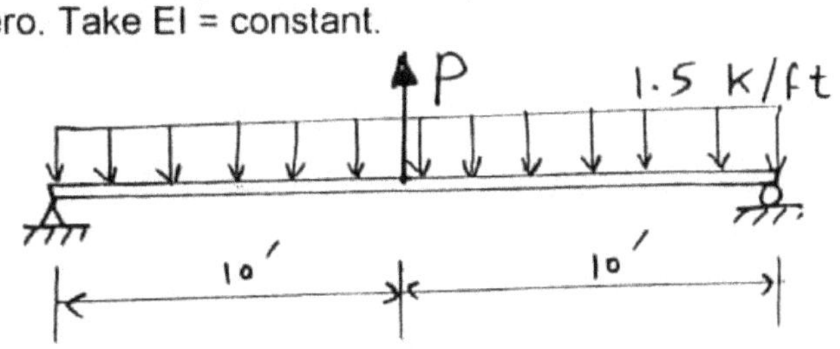

3- (20%) Vertical loads move along the bottom horizontal beam from A to C. Draw the influence lines for:

(a) the tension in the cable.
(b) the moment in the beam at B.
(c) the shear in the beam at E.

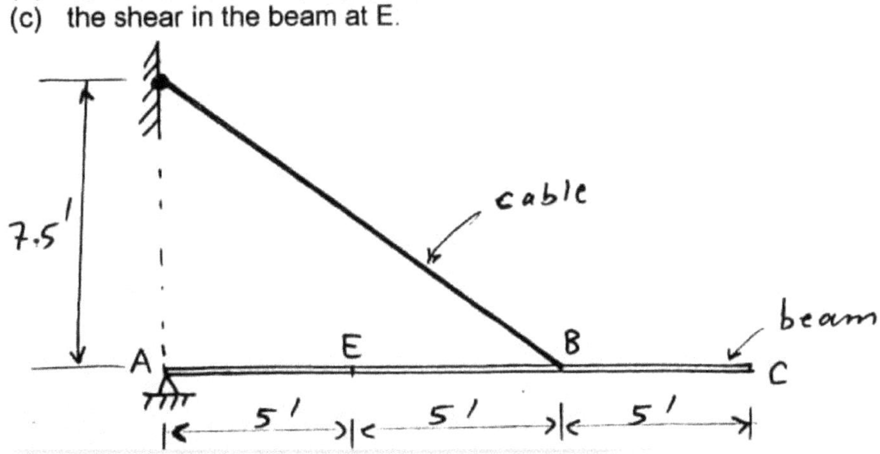

4- (20%) The three-hinged arch is composed of segments of a circle as shown in the figure. Supports A and E are not on the same level. Determine:

(a) the reactions at the supports.
(b) The shear, thrust, and moment in the arch just below point B.

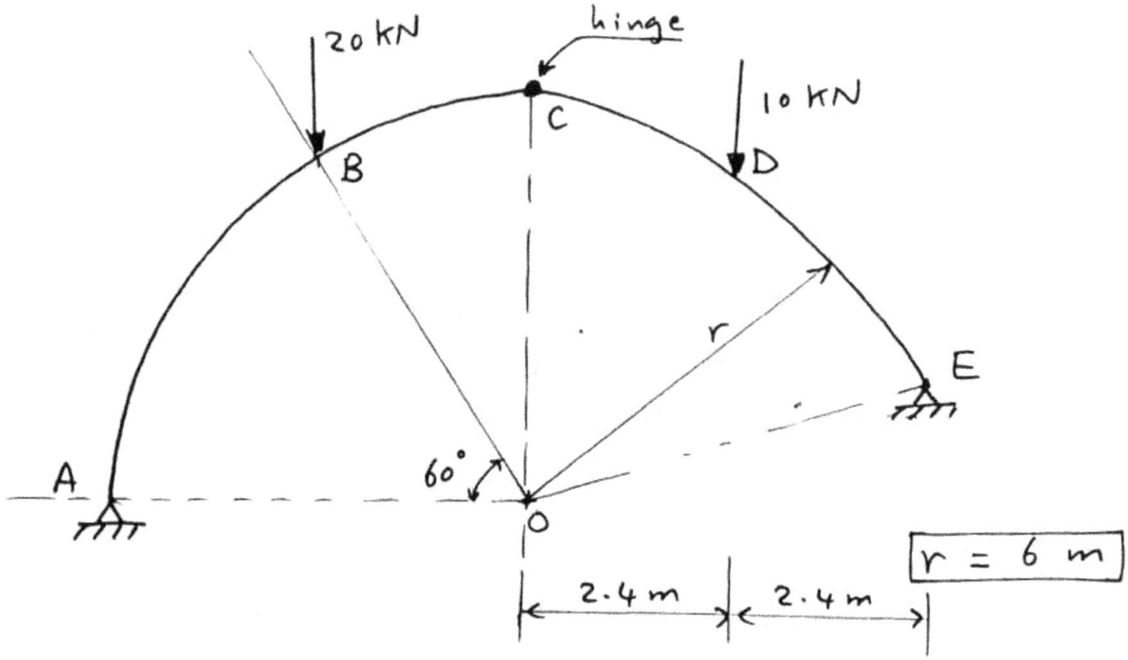

5. (30%) Draw the bending moment diagram for the redundant beam shown in the figure. Take EI = constant.

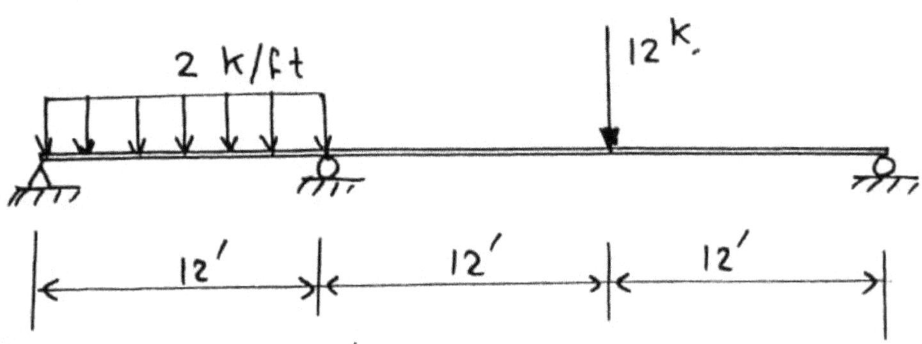

```
      ----------------------------------------------
        FORTRAN 77 Program for Analyzing Trusses
      ----------------------------------------------
      PROGRAM TRUSS
      DIMENSION A(50,50), P(50), F(50),C(50)
      PRINT*, 'Enter the number of simultaneous equations;'
      READ*,N
      PRINT*, 'Enter the number of members;'
      READ*,NM
      PRINT*, 'Enter the values of applied loads;'
      DO 5 I=1,N
        PRINT*, 'P(',I,'):'
        READ*, P(I)
        P(I)=-P(I)
    5 CONTINUE
      PRINT*, 'Enter the x and y coordinates at each joint'
      DO 10 I=1,N
        PRINT*, 'C(',I,'):'
        READ*, C(I)
   10 CONTINUE
      DO 15 M=1,NM
        PRINT*, 'Enter the vector numbers  i, j, k, l for member ', M
        READ*, I, J, K, L
        ALENGTH= SQRT(((C(K)-C(I))**2) + ((C(L)-C(J))**2))
        A(I,M)=(C(K)-C(I))/ALENGTH
        A(J,M)=(C(L)-C(J))/ALENGTH
        A(K,M)=-A(I,M)
        A(L,M)=-A(J,M)
   15 CONTINUE
      DO 20 M= (NM+1), N
        PRINT*, 'Enter the vector number for reaction ', M
        READ*, I
        A(I,M)=1.0
   20 CONTINUE
      CALL SIMUS(A,P,F,N)
      PRINT*, ' Solution:'
      PRINT*, ' ---------'
      DO 25 I=1,NM
        PRINT 40, ' Member    ', I, F(I)
   25 CONTINUE
      DO 30 I= (NM+1), N
        PRINT 40, ' Reaction   ', I, F(I)
   30 CONTINUE
   40 FORMAT(A ,1X,I2,F12.5)
      END

      SUBROUTINE SIMUS(A,B,X,N)
      DIMENSION A(50,50), B(50), X(50), IQ(50)
  000 FORMAT(/,1X, 'Coefficient matrix is singular')
      DO 10 I=1,N
   10 IQ(I)=I
      NN=N-1
      DO 55 I=1,NN
        I1=I+1
        PIVOT=ABS(A(I,I))
        IR=I
        IC=I
        DO 15 J=I,N
          DO 15 K=I,N
            AA=ABS(A(J,K))
            IF(AA.LE.PIVOT) GO TO 15
            PIVOT = AA
            IR=J
            IC=K
   15   CONTINUE
        IF(PIVOT.NE.0) GO TO 20
        PRINT 1000
        GO TO 100
   20   IF(IR.EQ.I) GO TO 30
        DO 25 J=I,N
          AA=A(I,J)
          A(I,J)=A(IR,J)
   25   A(IR,J)=AA
        AA=B(I)
        B(I)=B(IR)
        B(IR)=AA
   30   IF(IC.EQ.I) GO TO 40
        DO 35 J=1,N
          AA=A(J,I)
          A(J,I)=A(J,IC)
   35   A(J,IC)=AA
        K=IQ(I)
        IQ(I)=IQ(IC)
        IQ(IC)=K
   40   DO 50 J=I1,N
          C=A(J,I)/A(I,I)
          DO 45 K=I1,N
   45     A(J,K)=A(J,K)-C*A(I,K)
   50   B(J)=B(J)-C*B(I)
   55 CONTINUE
      X(N)=B(N)/A(N,N)
      DO 65 I1=1,NN
        I=N-I1
        X(I)=B(I)
        I1=I+1
        DO 60 J=I1,N
   60   X(I)=X(I)-A(I,J)*X(J)
   65 X(I)=X(I)/A(I,I)
      DO 75 I=1,N
        IF(IQ(I).EQ.I) GO TO 75
        I1=I+1
        DO 70 J=I1,N
          IF(IQ(J).NE.I) GO TO 70
          AA=X(I)
          X(I)=X(J)
          X(J)=AA
          IQ(J)=IQ(I)
          GO TO 75
   70   CONTINUE
   75 CONTINUE
  100 RETURN
      END
```

```
      PROGRAM SOLVE
C This program will solve a linear system A*X=B
      DIMENSION A(50,50),B(50)
      PRINT*, ' Enter the number of equations:'
      READ*, N
      IF(N.EQ.0) STOP
      PRINT*, 'Enter the linear system one equation at a time:'
      PRINT*, 'Conclude each equation with its right-hand constant:'
      DO 20 I=1,N
        PRINT*, 'Enter coefficients of equation', I
   20 READ*, (A(I,J),J=1,N), B(I)
C Solve the linear system
      CALL LINSYS(A,B,N,50,IER)
      PRINT 1000, N, IER
 1000 FORMAT(///, ' N=',I2,5X,'IER=',I3,//,'  I           Solution',/)
      PRINT 1001, (I,B(I),I=1,N)
 1001 FORMAT(I3,1PE20.10)
      STOP
      END
C
      SUBROUTINE LINSYS(MAT,B,N,MD,IER)
C This subroutine solves a system of linear equations
C            A*X=B
C The method used is Gaussian elimination with partial pivoting
C Input:
C The coefficient matrix A is stored in the array MAT
C The right-side constants are stored in the array B
C The order of the linear system is N (number of equations)
C The variable MD is the number of rows that MAT is dimensioned
C as having in the calling program
C Output:
C The array B contains the solution X
C MAT contains the upper triangular matrix U obtained by elimination
C The row multipliers used in the elimination are stored in
C the lower triangular part of MAT
C IER=0 means the matrix A was computationally nonsingular
C and the Gaussian elimination was completed satisfactorily.
C IER=1 means that the matrix A was computationally singular
      INTEGER PIVOT
      REAL MULT,MAT
      DIMENSION MAT(MD,*),B(*),PIVOT(100)
C Begin eliminations steps
      DO 40 K=1,N-1
C Choose pivot row
        PIVOT(K)=K
        AMAX=ABS(MAT(K,K))
        DO 10 I=K+1,N
          ABSA=ABS(MAT(I,K))
          IF(ABSA.GT.AMAX) THEN
            PIVOT(K)=I
            AMAX=ABSA
          END IF
   10   CONTINUE
        IF(AMAX.EQ.0.0) THEN
C Coefficient matrix is singular
          IER=1
          RETURN
        END IF
        IF(PIVOT(K).NE.K) THEN
C Switch rows K and PIVOT(K)
          I=PIVOT(K)
          TEMP=B(K)
          B(K)=B(I)
          B(I)=TEMP
          DO 20 J=K,N
            TEMP=MAT(K,J)
            MAT(K,J)=MAT(I,J)
   20       MAT(I,J)=TEMP
        END IF
C Perform step #K of elimination
        DO 30 I=K+1,N
          MULT=MAT(I,K)/MAT(K,K)
          MAT(I,K)=MULT
          B(I)=B(I)-MULT*B(K)
          DO 30 J=K+1,N
   30     MAT(I,J)=MAT(I,J)-MULT*MAT(K,J)
   40 CONTINUE
      IF(MAT(N,N).EQ.0.0) THEN
C Coefficient matrix is singular
        IER=1
        RETURN
      END IF
C Solve for solution X using back substitution
      DO 60 I=N,1,-1
        SUM=0.0
        DO 50 J=I+1,N
   50   SUM=SUM+MAT(I,J)*B(J)
   60 B(I)=(B(I)-SUM)/MAT(I,I)
      IER=0
      RETURN
      END
```

Determinate and Indeterminate Structures

* Why do we build indeterminate structures ?

Example-1- : Trusses :

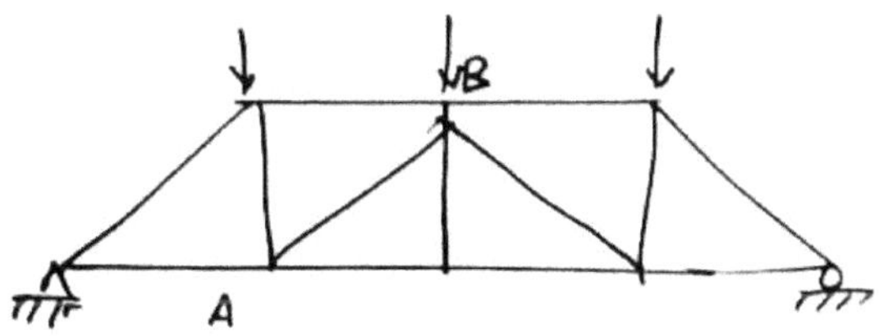

determinate

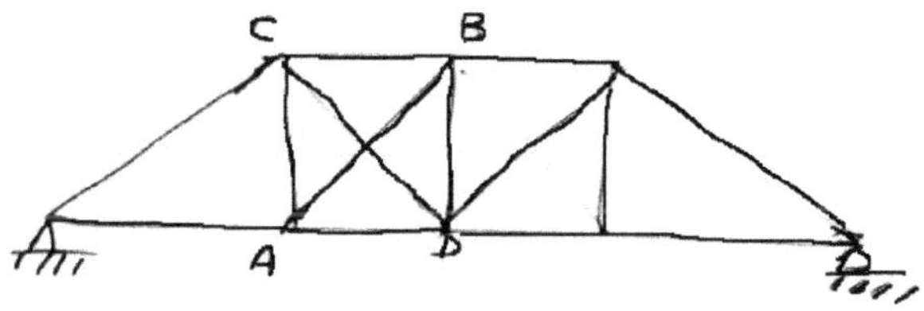

indeterminate (internally)

If member AB fails, the determinate truss will become <u>unstable</u> leading to <u>collapse</u> of the structure.

If member AB fails, the indeterminate truss will still be <u>stable</u> due to the presence of member CD.

---

(1) There is a <u>risk factor</u> in building determinate structure.

(2) Indeterminate structures are more economical (no special construction techniques for rollers and internal hinges).

Example-2-: Frames :

With determinate frames, we can only build one-storey buildings.

All multi-storey buildings are indeterminate frames. (multi-bay and multi-storey frames)

determinate (one storey) — indeterminate — indeterminate — indeterminate

Problem 5.42 :

(H11)

Find x so that the maximum moment in span ABC is equal to the maximum moment in span CD.

Solution :

The problem is symmetric.
It is determinate.

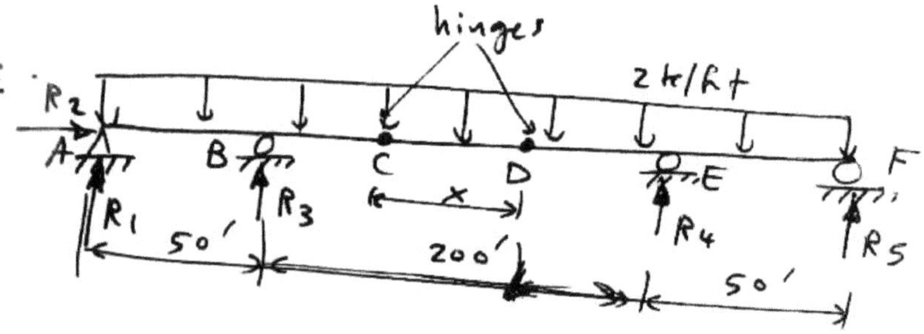

(1) Calculate the reactions :

$\xrightarrow{+} \Sigma F_x = 0 : \; R_2 = 0.$

Segment CD :

From symmetry,

$$R_C = R_D = \frac{2x}{2} = x, \text{ k}.$$

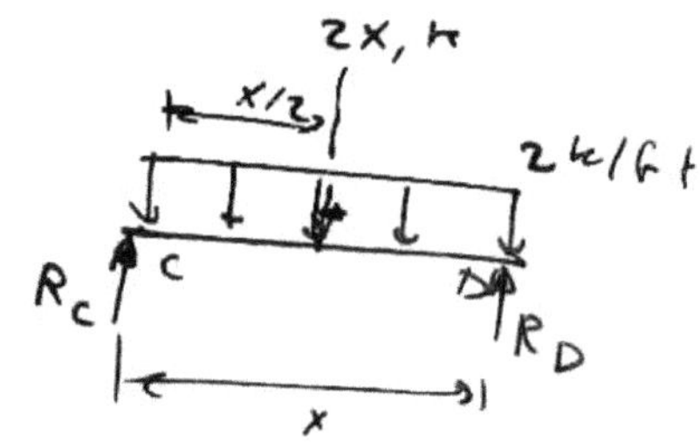

Segment ABC :

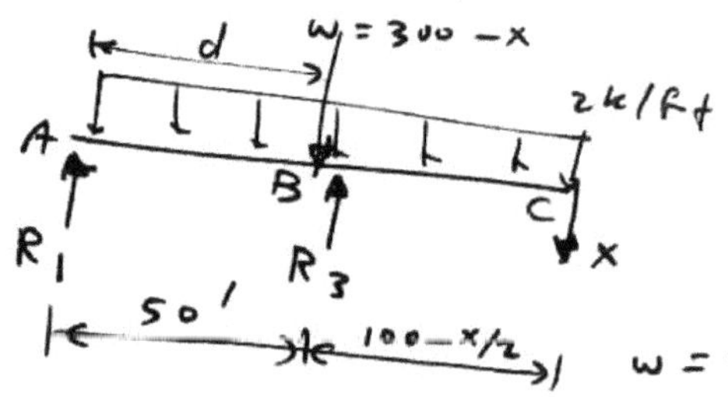

$\circlearrowleft^{+} \Sigma M_A = 0 :$

$$R_3(50) - x\left(150 - \frac{x}{2}\right) - (300 - x)\left(75 - \frac{x}{4}\right) = 0$$

$$\therefore 50R_3 - 150x + \frac{1}{2}x^2 - 22{,}500 + 75x + 75x - \frac{1}{4}x^2 = 0$$

$$w = 2\left(50 + 100 - \frac{x}{2}\right) = 2\left(150 - \frac{x}{2}\right) = 300 - x, \text{ k}.$$

$$d = \frac{50 + 100 - x/2}{2} = 75 - x/4.$$

$$\therefore 50R_3 - 22{,}500 + \frac{1}{4}x^2 = 0$$

$$\therefore 50R_3 = 22{,}500 - \frac{1}{4}x^2$$

$$R_3 = \frac{90{,}000 - x^2}{200} \quad \checkmark$$

$+\uparrow \Sigma F_y = 0 :$

$$R_1 + \frac{90{,}000 - x^2}{200} - x - (300 - x) = 0$$

$$R_1 = \frac{x^2 - 90{,}000}{200} + 300 = \frac{x^2 - 30{,}000}{200}.$$

From symmetry of the entire beam:

$$R_4 = R_3 = \frac{90{,}000 - x^2}{200}$$

$$R_5 = R_1 = \frac{x^2 - 30{,}000}{200}$$

S.F.D.

$$V_B^- = \frac{x^2 - 30{,}000}{200} - 2(50) = \frac{x^2 - 50{,}000}{200}$$

$$V_B^+ = \frac{x^2 - 50{,}000}{200} + \frac{90{,}000 - x^2}{200} = 200\text{ k.}$$

$$V_C = 200 - 2\left(100 - \frac{x}{2}\right) = x.$$

$$V_D = x - 2x = -x.$$

$$V_E^- = -x - 2\left(100 - \frac{x}{2}\right) = -200\text{ k.}$$

$$V_E^+ = -200 + \frac{90{,}000 - x^2}{200} = \frac{50{,}000 - x^2}{200}$$

$$V_F = \frac{50{,}000 - x^2}{200} - 2(50) = \frac{30{,}000 - x^2}{200}$$

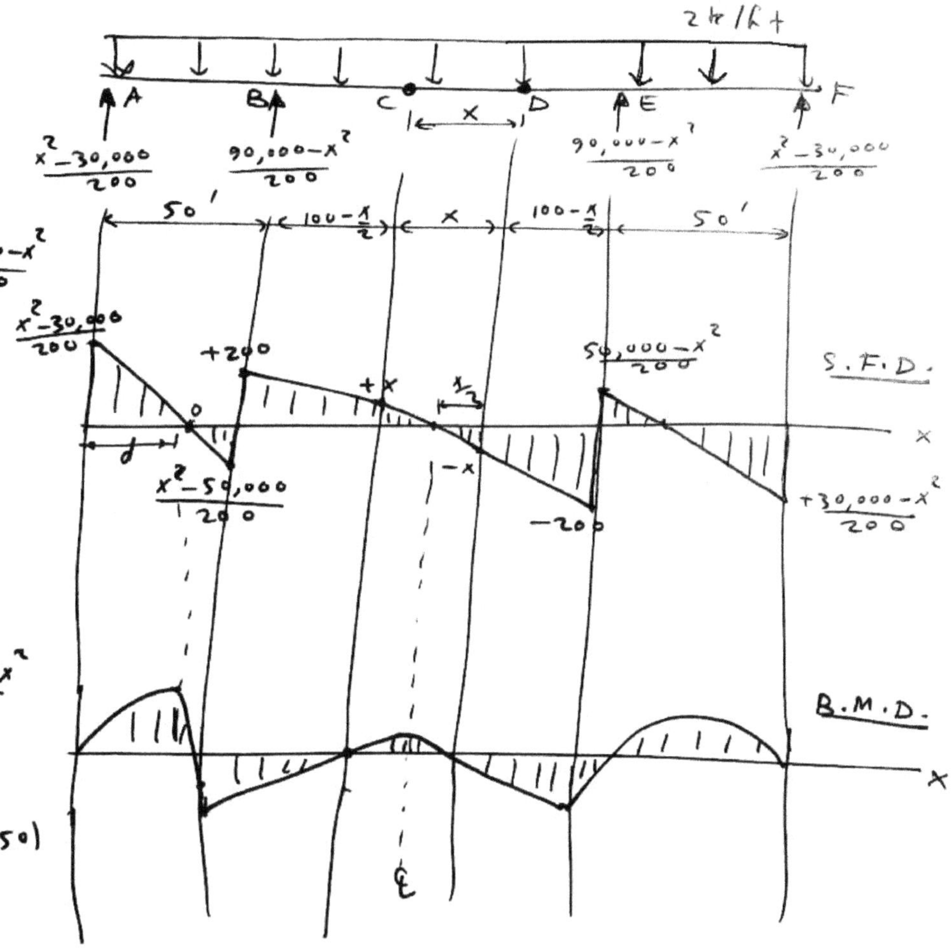

Locate point 0 of zero shear:

$$\frac{x^2 - 30{,}000}{200d} = \frac{x^2 - 50{,}000}{200(50-d)}$$

$$200d(x^2 - 50{,}000) = 200(50)(x^2 - 30{,}000) - 200d(x^2 - 30{,}000)$$

$$\cancel{200}d\left[x^2 - 50{,}000 + x^2 - 30{,}000\right] = \cancel{200}(50)(x^2 - 30{,}000)$$

$$d(2x^2 - 80{,}000) = 50(x^2 - 30{,}000)$$

$$2d(x^2 - 40{,}000) = 50(x^2 - 30{,}000) \Longrightarrow d = \frac{25(x^2 - 30{,}000)}{(x^2 - 40{,}000)}$$

D.M.D. (H13)

$$M_{o(max)} = \frac{1}{2}\left(\frac{x^2-30,000}{200}\right)25\left(\frac{x^2-30,000}{x^2-40,000}\right)$$

$$= \frac{25}{400}\frac{(x^2-30,000)^2}{(x^2-40,000)}$$ (Maximum moment in span ABC).

In span CD, the maximum moment lies in the middle (from symmetry).

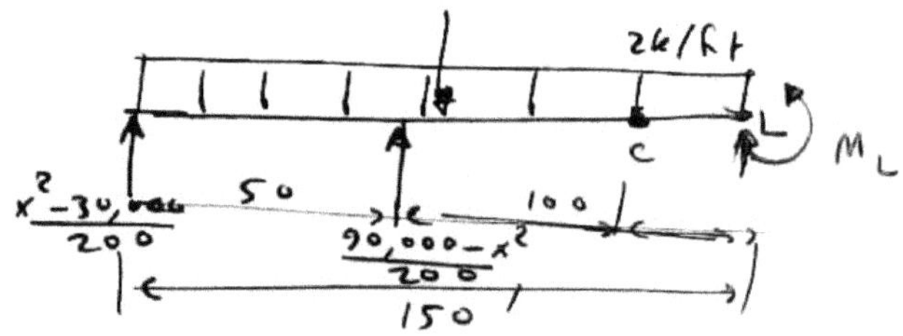

$\circlearrowleft + \Sigma M_L = 0$

$$M_L + 300\left(\frac{150}{2}\right) - \left(\frac{90,000-x^2}{200}\right)(100) - \left(\frac{x^2-30,000}{200}\right)(150) = 0$$

$$\therefore M_L = \frac{150}{200}\left(x^2-30,000\right) + \frac{100}{200}\left(90,000-x^2\right) - 22,500$$

$$= \left(\frac{150x^2}{200} - \frac{100x^2}{200}\right) - \frac{150}{200}(30,000) + \frac{100}{200}(90,000) - 22,500$$

$M_L = 0.25x^2$ (max. moment in span CD).

Equate the two maximum moments:

$$\frac{25}{400}\frac{(x^2-30,000)^2}{x^2-40,000} = 0.25x^2$$

$$(x^2-30,000)^2 = 4x^2(x^2-40,000)$$

$$x^4 - 60,000x^2 + (30,000)^2 = 4x^4 - 160,000x^2$$

$$3x^4 - 100,000x^2 - (30,000)^2 = 0$$

$$\therefore x^2 = \frac{100,000 \pm \sqrt{(100,000)^2 - 4(3)(-(30,000)^2)}}{2(3)} = \frac{100,000 \pm 144,222}{6}$$

$$x^2 = 40{,}704 \Longrightarrow x = 201.8 \text{ ft.}$$

(H14)

Check the maximum moment at B:

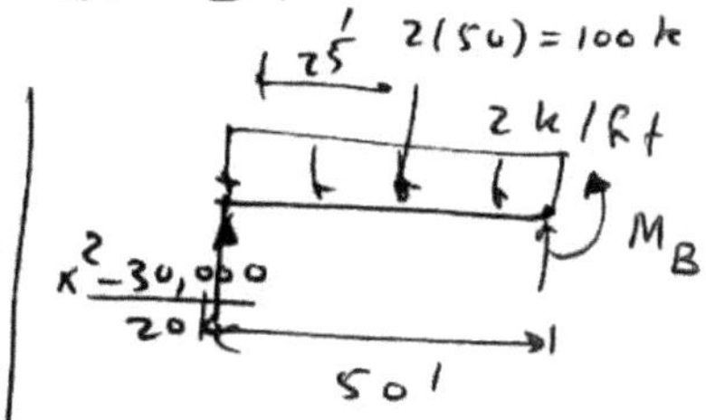

$+\circlearrowleft \Sigma M_B = 0$:

$$M_B + 100(25) - \left(\frac{x^2-30{,}000}{200}\right)(50) = 0$$

$$M_B = \frac{x^2-30{,}000}{4} - 2500$$

$$= \frac{x^2-40{,}000}{4}.$$

Equate it to M_L: $\frac{x^2-40{,}000}{4} = 0.25x^2$ (impossible).

Homework 3

Problem 4.2 :

Determine the reactions and all the bar forces.

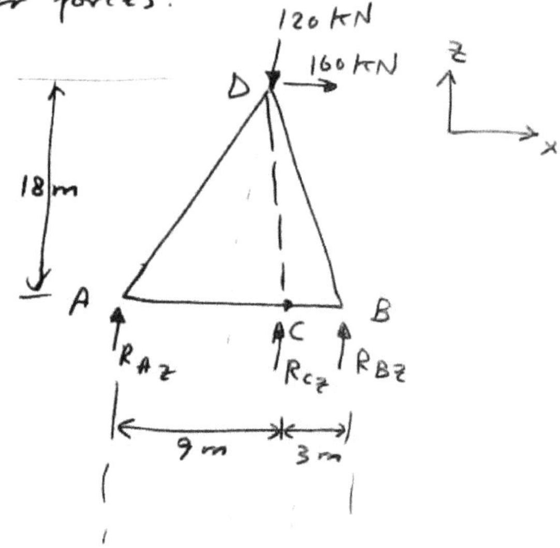

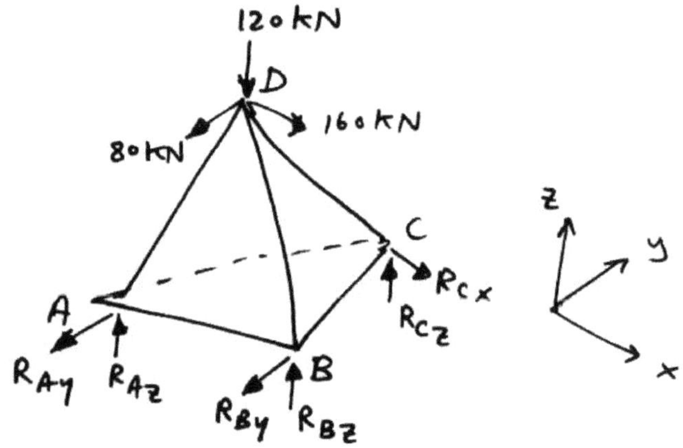

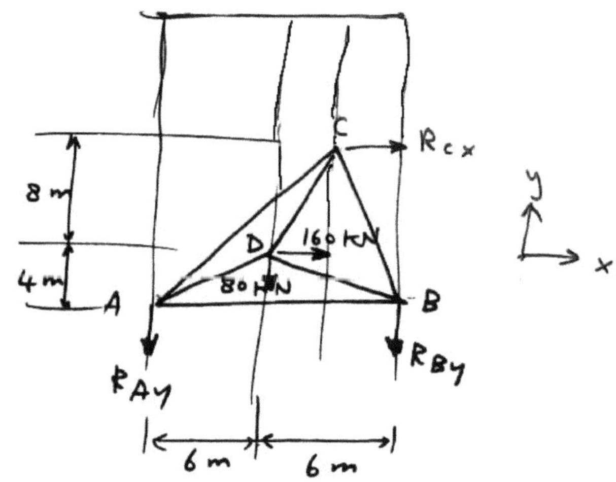

Solution :

there are six reactions.

there are six equations of equilibrium.

∴ The truss is externally determinate.

No. of joints $j = 4$

3 equations at each joint.

∴ $3j = 3(4) = 12$ equations.

No. of unknowns: members $m = 6$, reactions $r = 6$

∴ $m + r = 12$.

∴ $3j = m + r$.

∴ the truss is determinate.

(1) Calculate the reactions :

$$\Sigma M_{x(AB)} = 0 \quad : R_{Cz}(12) - 120(4) + 80(18) = 0$$

$$\Rightarrow R_{Cz} = -80 \text{ kN} \Rightarrow R_{Cz} = 80 \text{ kN} \downarrow$$

$\Sigma M_{y(B)} = 0$:

$\overset{-80}{R_{cz}}(3) + R_{Az}(12) + 160(18) - 120(6) = 0$

$\Rightarrow R_{Az} = -160 \text{ kN} \Rightarrow R_{Az} = 160 \text{ kN} \downarrow$

$\Sigma F_z = 0$: $R_{Az} + R_{Bz} + R_{cz} - 120 = 0$

$-160 + R_{Bz} - 80 - 120 = 0$

$\Rightarrow R_{Bz} = 360 \text{ kN} \uparrow$

$\Sigma F_x = 0$: $R_{cx} + 160 = 0 \Rightarrow R_{cx} = -160 \text{ kN}$

$\therefore R_{cx} = 160 \text{ kN} \nwarrow$

$\Sigma M_{z(B)} = 0$: $-\overset{-160}{R_{cx}}(12) + R_{Ay}(12) + 80(6) - 160(4) = 0$

$\Rightarrow R_{Ay} = -146.7 \text{ kN}$

$\Rightarrow R_{Ay} = 146.7 \text{ kN} \nearrow$

$\Sigma F_y = 0$: $-\overset{-146.7}{R_{Ay}} - R_{By} - 80 = 0 \Rightarrow R_{By} = 66.7 \text{ kN} \swarrow$

Member	Projection			Length	Ratio of projection to length (Direction Cosines)		
	x	y	z	L	x/L	y/L	z/L
AB	12	0	0	12	1.0	0	0
AC	9	12	0	15	0.6	0.8	0
BC	3	12	0	12.37	0.243	0.970	0
AD	6	4	18	19.39	0.309	0.206	0.928
BD	6	4	18	19.39	0.309	0.206	0.928
CD	3	8	18	19.92	0.151	0.402	0.904

(2) Use the method of joints :

Joint A :

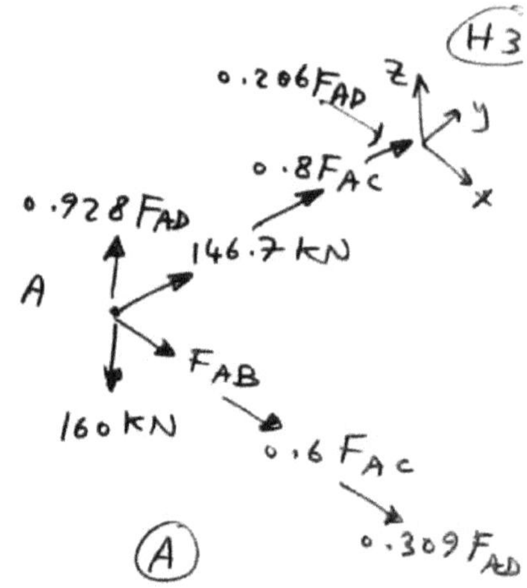

$\Sigma F_x = 0$:

$F_{AB} + 0.6\, F_{AC} + 0.309\, F_{AD} = 0$ —— (1)

$\Sigma F_y = 0$:

$146.7 + 0.8 F_{AC} + 0.206\, F_{AD} = 0$ —— (2)

$\Sigma F_z = 0$:

$0.928\, F_{AD} - 160 = 0$

$\Rightarrow F_{AD} = 172.4$ kN (T).

Substitute in eq.(2) :

$146.7 + 0.8\, F_{AC} + 0.206\,(172.4) = 0$

$\Rightarrow F_{AC} = -227.8$ kN

$\therefore F_{AC} = 227.8$ kN (C).

Substitute in eq. (1) :

$F_{AB} + 0.6\,(-227.8) + 0.309\,(172.4) = 0$

$\Rightarrow F_{AB} = 83.4$ kN (T).

Joint B :

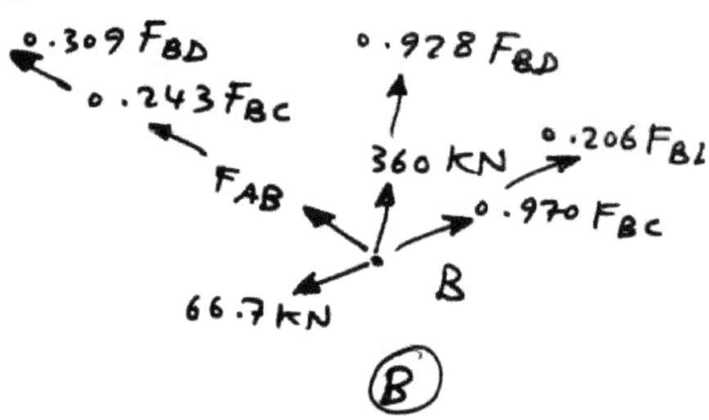

$\Sigma F_z = 0$:

$0.928\, F_{BD} + 360 = 0$

$\Rightarrow F_{BD} = -387.9$ kN.

$\therefore F_{BD} = 387.9$ kN (C).

$\Sigma F_x = 0$:

$-F_{AB} - 0.243\, F_{BC} - 0.309\, F_{BD} = 0$.

$-83.4 - 0.243\, F_{BC} - 0.309\,(-387.9) = 0$

$\Rightarrow F_{BC} = 150.0$ kN (T).

Check: $\Sigma F_y = 0$:

$$0.970\,F_{BC} + 0.206\,F_{BD} - 66.7 = 0$$

$$0.970\,(150) + 0.206\,(-387.9) - 66.7 = 0$$

$$-1.1 = 0 \quad \checkmark$$

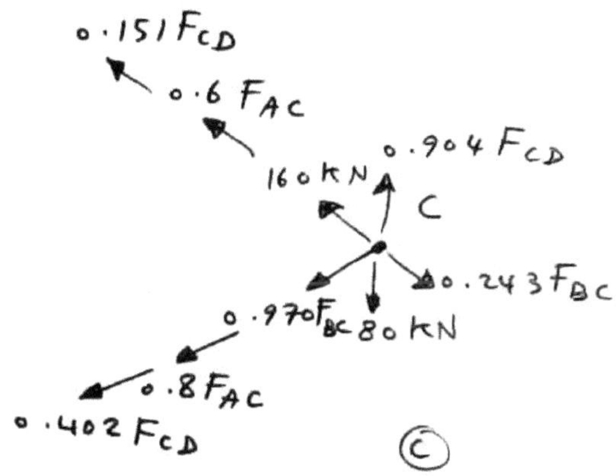

Joint C:

$\Sigma F_z = 0$:

$$0.904\,F_{CD} - 80 = 0$$

$$\Rightarrow \quad F_{CD} = 88.5 \text{ kN (T)}.$$

Check:

$\Sigma F_x = 0$:

$$-160 - 0.6\,F_{AC} - 0.151\,F_{CD} + 0.243\,F_{BC} = 0$$

$$-160 - 0.6\,(-227.8) - 0.151\,(88.5) + 0.243\,(150) = 0$$

$$-0.2 = 0 \quad \checkmark$$

$\Sigma F_y = 0$:

$$-0.970\,F_{BC} - 0.8\,F_{AC} - 0.402\,F_{CD} = 0$$

$$-0.970\,(150) - 0.8\,(-227.8) - 0.402\,(88.5) = 0$$

$$1.2 = 0 \quad \checkmark$$

Final Check at joint D:

$\Sigma F_x = 0$:

$$160 + 0.309\,F_{BD} + 0.151\,F_{CD} - 0.309\,F_{AD} = 0$$

$$160 + 0.309\,(-387.9) + 0.151\,(88.5) - 0.309\,(172.4) = 0$$

$$0.2 = 0 \quad \checkmark$$

$\Sigma F_y = 0$:

$$0.206\,F_{BD} + 0.206\,F_{AD} + 80 - 0.402\,F_{CD} = 0.$$

$$0.206\,(-387.9) + 0.206\,(172.4) + 80 - 0.402\,(88.5) = 0$$

$$0.03 = 0 \quad \checkmark$$

$\Sigma F_z = 0$: $-120 - 0.928\,F_{AD} - 0.928\,F_{BD} - 0.904\,F_{CD} = 0$

$$-120 - 0.928\,(172.4) - 0.928\,(-387.9) - 0.904\,(88.5) = 0.$$

$$0.02 = 0.$$

(145)

Problem 4.4 :

∴ Determine the reactions and all bar forces.

Solution :

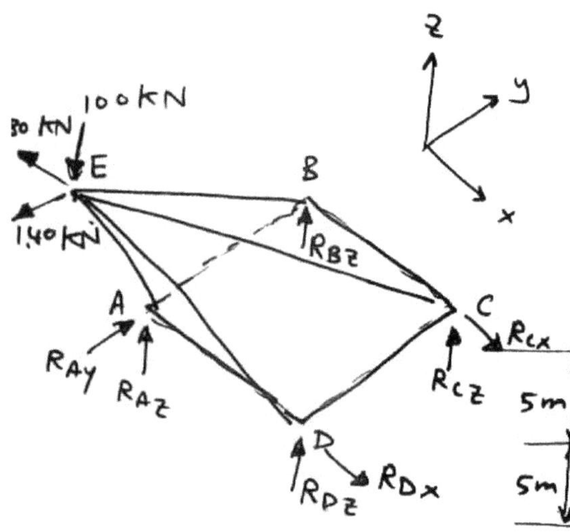

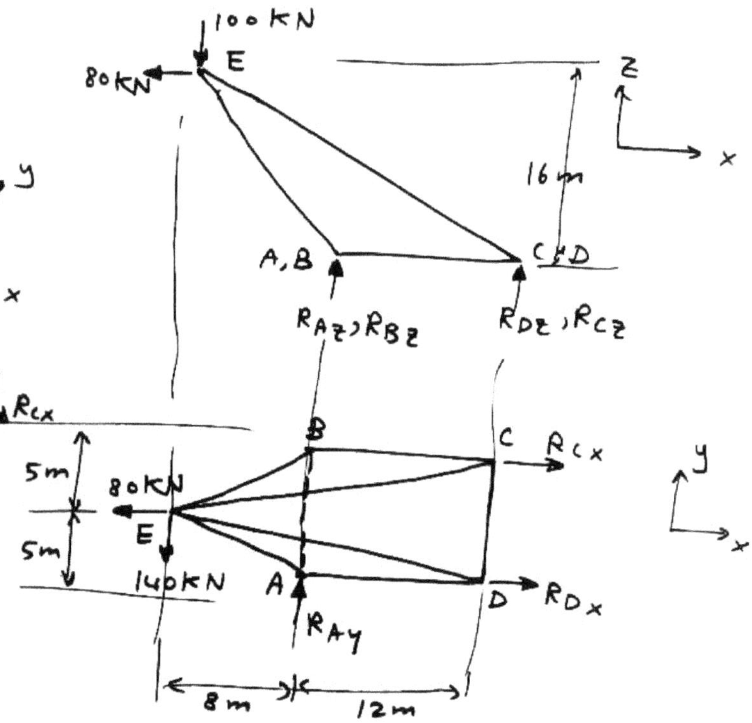

There are five joints, $j = 5$

∴ $3j = 15$ equations.

\# unknowns: members $m = 8$; reactions $r = 7$ } $m + r = 15$ unknown.

∴ $3j = m + r$.

∴ The truss is determinate.

Member	Projection x	Projection y	Projection z	Length L	Ratio of projection to length (Direction cosine) X/L	Y/L	Z/L
AB	0	10	0	10	0	1	0
AD	12	0	0	12	1	0	0
CD	0	10	0	10	0	1	0
BC	12	0	0	12	1	0	0
AE	8	5	16	18.57	0.431	0.269	0.862
BE	8	5	16	18.57	0.431	0.269	0.862
CE	20	5	16	26.10	0.766	0.192	0.613
DE	20	5	16	26.10	0.766	0.192	0.613

Joint A:

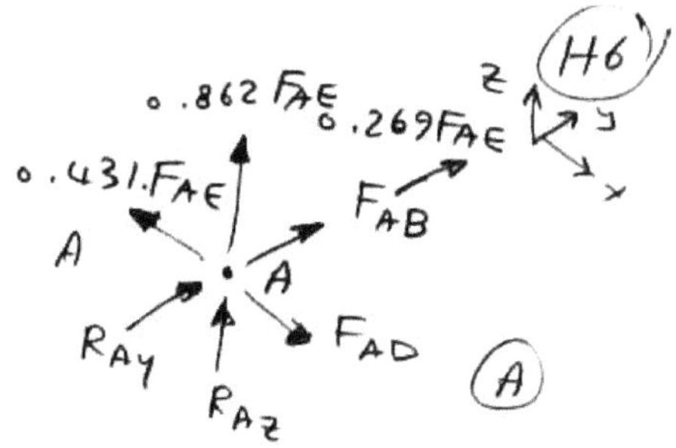

$\Sigma F_x = 0$:

$F_{AD} - 0.431\,F_{AE} = 0$ ——— (1)

$\Sigma F_y = 0$:

$F_{AB} + 0.269\,F_{AE} + R_{Ay} = 0$ ——— (2)

$\Sigma F_z = 0$:

$R_{Az} + 0.862\,F_{AE} = 0$ ——— (3)

Joint B:

$\Sigma F_x = 0$:

$F_{BC} - 0.431\,F_{BE} = 0$ ——— (4)

$\Sigma F_y = 0$:

$-F_{AB} - 0.269\,F_{BE} = 0$ ——— (5)

$\Sigma F_z = 0$:

$R_{Bz} + 0.862\,F_{BE} = 0$ ——— (6)

Joint C:

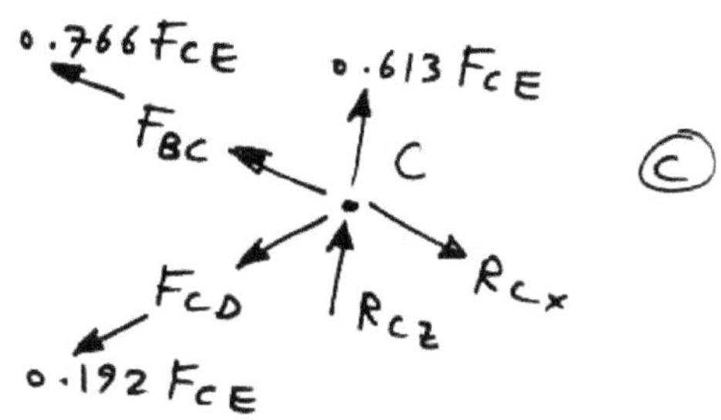

$\Sigma F_x = 0$:

$R_{Cx} - F_{BC} - 0.766\,F_{CE} = 0$ ——— (7)

$\Sigma F_y = 0$:

$-F_{CD} - 0.192\,F_{CE} = 0$ ——— (8)

$\Sigma F_z = 0$:

$R_{Cz} + 0.613\,F_{CE} = 0$ ——— (9)

Joint D:

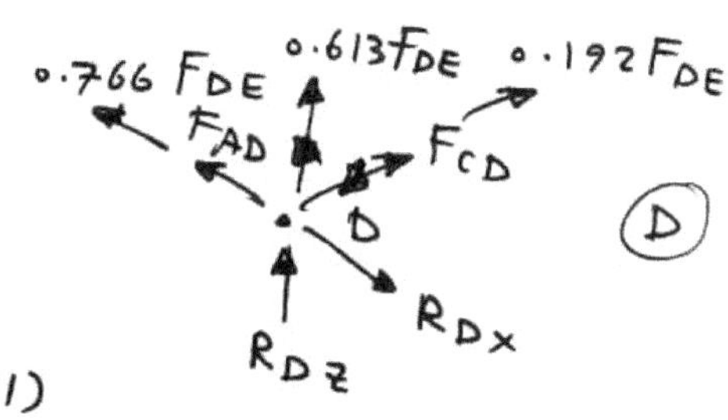

$\Sigma F_x = 0$:

$R_{Dx} - F_{AD} - 0.766\,F_{DE} = 0$ ——— (10)

$\Sigma F_y = 0$:

$F_{CD} + 0.192\,F_{DE} = 0$ ——— (11)

$\Sigma F_z = 0$:

$R_{Dz} + 0.613\,F_{DE} = 0$ ——— (12)

(H7)

Joint E :

$\Sigma F_x = 0$:

$0.431 F_{AE} + 0.431 F_{BE} + 0.766 F_{CE}$
$+ 0.766 F_{DE} - 80 = 0$ ———— (13)

$\Sigma F_y = 0$:

$0.269 F_{BE} + 0.192 F_{CE} - 140$
$- 0.269 F_{AE} - 0.192 F_{DE} = 0$ ———— (14)

$\Sigma F_z = 0$:

$-100 - 0.862 F_{AE} - 0.862 F_{BE}$
$- 0.613 F_{CE} - 0.613 F_{DE} = 0$ ———— (15)

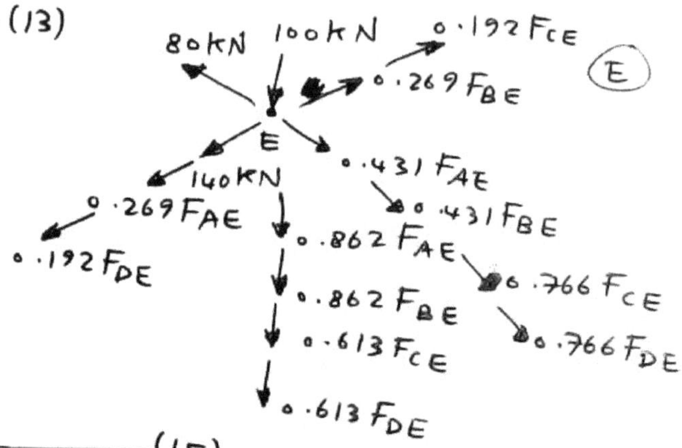

* Write the equations in <u>matrix</u> form :

	F_{AB}	F_{AD}	F_{CD}	F_{BC}	F_{AE}	F_{BE}	F_{CE}	F_{DE}	R_{Az}	R_{Bz}	R_{Cz}	R_{Dz}	R_{Cx}	R_{Dx}	R_{Ay}
1	0	1	0	0	$\overline{0}.431$	0	0	0	0	0	0	0	0	0	0
2	1	0	0	0	0.269	0	0	0	0	0	0	0	0	0	1
3	0	0	0	0	0.862	0	0	0	1	0	0	0	0	0	0
4	0	0	0	1	0	$\overline{0}.431$	0	0	0	0	0	0	0	0	0
5	−1	0	0	0	0	$\overline{0}.269$	0	0	0	0	0	0	0	0	0
6	0	0	0	0	0	0.862	0	0	0	1	0	0	0	0	0
7	0	0	0	−1	0	0	$\overline{0}.766$	0	0	0	0	0	1	0	0
8	0	0	−1	0	0	0	$\overline{0}.192$	0	0	0	0	0	0	0	0
9	0	0	0	0	0	0	0.613	0	0	0	1	0	0	0	0
10	0	−1	0	0	0	0	0	$\overline{0}.766$	0	0	0	0	0	1	0
11	0	0	1	0	0	0	0	0.192	0	0	0	0	0	0	0
12	0	0	0	0	0	0	0	0.613	0	0	0	1	0	0	0
13	0	0	0	0	0.431	0.431	0.766	0.766	0	0	0	0	0	0	0
14	0	0	0	0	$\overline{0}.269$	0.269	0.192	$\overline{0}.192$	0	0	0	0	0	0	0
15	0	0	0	0	$\overline{0}.862$	→ −0.862	$\overline{0}.613$	$\overline{0}.613$	0	0	0	0	0	0	0

$\times \{F_{AB}, F_{AD}, F_{CD}, F_{BC}, F_{AE}, F_{BE}, F_{CE}, F_{DE}, R_{Az}, R_{Bz}, R_{Cz}, R_{Dz}, R_{Cx}, R_{Dx}, R_{Ay}\}^T = \{0, 0, 0, 0, 0, 0, 0, 0, 0, 0, 0, 0, 80, 140, 100\}^T$

Solving the system of equations using the computer, we obtain the following results: (H8)

$F_{AB} = -27.336$ kN	$R_{Az} = 361.0$ kN
$F_{AD} = -180.51$ kN	$R_{Bz} = -87.6$ kN
$F_{CD} = -27.160$ kN	$R_{Cz} = -86.71$ kN
$F_{BC} = 43.80$ kN	$R_{Dz} = -86.71$ kN
$F_{AE} = -418.8$ kN	$R_{Cx} = 152.2$ kN
$F_{BE} = 101.6$ kN	$R_{Dx} = -72.16$ kN
$F_{CE} = 141.5$ kN	$R_{Ay} = 140.0$ kN
$F_{DE} = 141.5$ kN	

Homework 8 (H1)

①

$$\Delta R = (\Sigma P)\, d/L + P' e/L - P_1$$

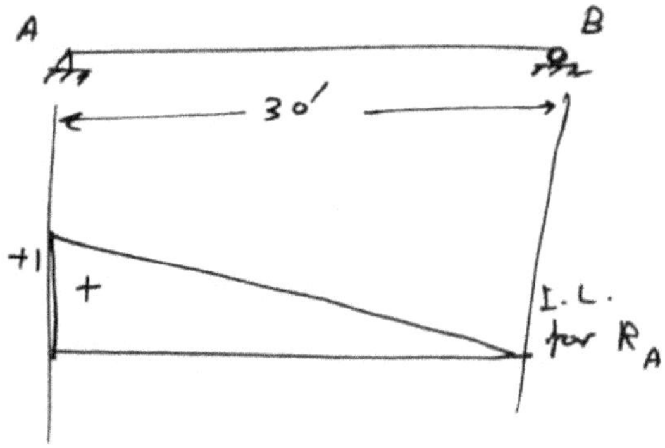

Move the wheels to the left until wheel ① is placed over A, then move further to the left until wheel ② comes on A, thus:

$$\Delta R_{2-1} = [40+40+40+30]\left(\frac{5}{30}\right) + 30\left(\frac{4}{30}\right) - 20$$

$$= +9 > 0 \text{ (an increase).}$$

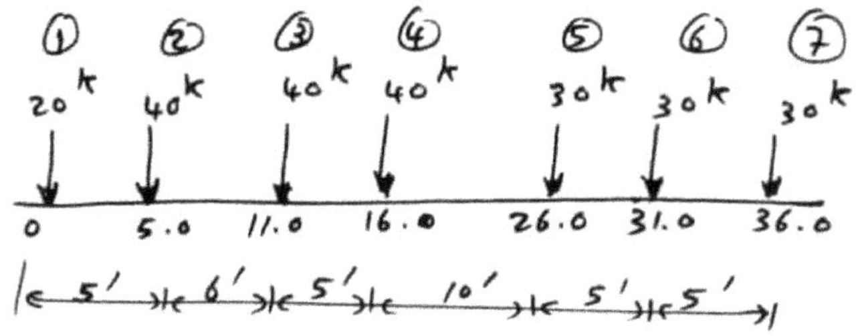

②-①: $e' = 35 - 31 = 4'$.

③-②: $e' = 41 - 36 = 5'$.

Again, move the wheels to the left from the previous position until wheel ③ comes on support A, thus:

$$\Delta R_{3-2} = (40+40+30+30)\left(\frac{6}{30}\right) + 30\left(\frac{5}{30}\right) - 40$$

$$= -7 < 0 \quad \text{(a decrease).}$$

Since the reaction decreases by moving from wheel ② on A to wheel ③ on A, the last movement should not be made, and the former position of the wheels (wheel ② on A) is the one producing maximum R_A.

$$\therefore R_{A(max)} = \frac{\Sigma (P \cdot x)}{L}$$

$$= [30(4) + 30(9) + 40(19) + 40(24) + 40(30)]/30$$

$$= \underline{\underline{+110.3\ k.}}$$

(H2)

② $\Delta V = (\Sigma P)d/L + P'e'/L + P''y'' - P_1$

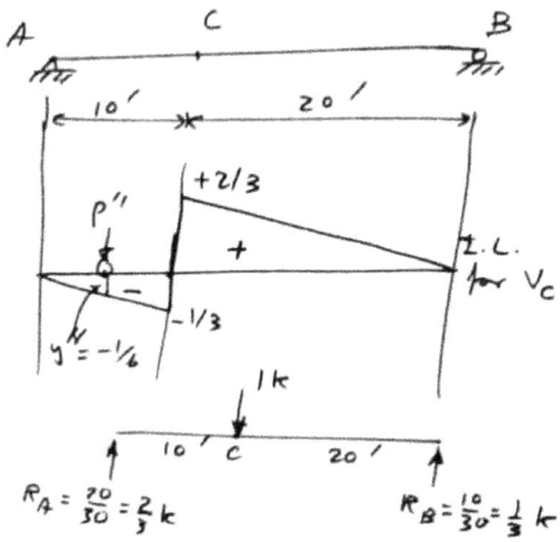

Wheel ① at section C to wheel ② at the section:

$\Delta V_{2-1} = [20 + 40 + 40 + 40]\left(\frac{5}{30}\right) - 20$

$= +3.3 > 0$ (an increase).

Wheel ② at section C to wheel ③ at the section:

$\Delta V_{3-2} = [40 + 40 + 40]\left(\frac{6}{30}\right)$
$+ 30\left(\frac{5}{30}\right) + 30\left(\frac{0}{30}\right) + 20\left(\frac{1}{6}\right) - 40$

$= -7.67 < 0$ (a decrease).

Therefore, wheel ② should be placed at the section for maximum positive shear.

Max. Positive $V_c = \frac{\Sigma(P \cdot x)}{L} - \Sigma P_L$

$= [40(9) + 40(14) + 40(20) + 20(25)]/30$
$- 20$
$= \underline{\underline{+54.0}}$ k.

---

③ Try wheel ② placed at:

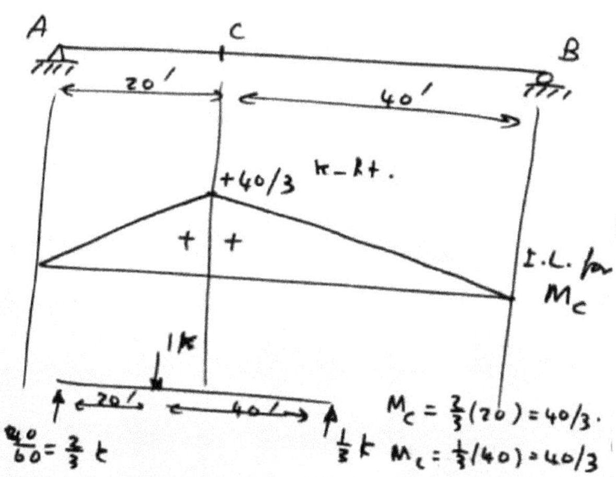

just to the right of C:

$W_L = 10$ k

$W_R = 30 + 20 + 20 + 30 + 20 = 120$ k.

$\frac{W_R}{b} = \frac{120}{40} = 3 > \frac{W_L}{a} = \frac{10}{20} = 0.5$ ✓

just to the left of C:

$W_L = 10 + 30 = 40$ k

$W_R = 120 - 30 = 90$ k.

(H3)

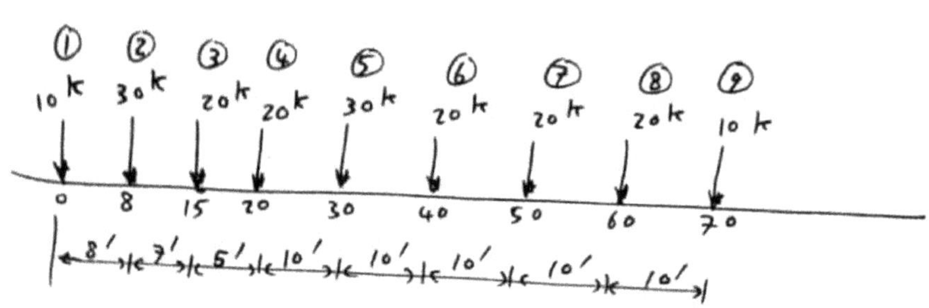

$\therefore \quad \frac{W_R}{b} = \frac{90}{40} = 2.25 \not< \frac{W_L}{a} = \frac{40}{20} = 2.0$.

$\therefore$ wheel ② does not satisfy the criterion.

Try wheel ③ placed at :

just to the right of C :

$W_L = 10 + 30 = 40$ k .

$W_R = 20 + 20 + 30 + 20 + 20 = 110$ k .

$\frac{W_R}{b} = \frac{110}{40} = 2.75 > \frac{W_L}{a} = \frac{40}{20} = 2.0$ ✓

just to the left of C :

$W_L = 10 + 30 + 20 = 60$ k .

$W_R = 110 - 20 = 90$ k .

$\frac{W_R}{b} = \frac{90}{40} = 2.25 < \frac{W_L}{a} = \frac{60}{20} = 3.0$ ✓ .

$\therefore$ wheel ③ satisfies the criterion.

Try wheel ④ placed at :

Just to the right of C :

$W_L = 10 + 30 + 20 = 60$ k .

$W_R = 20 + 30 + 20 + 20 + 20 = 110$ k .

$\frac{W_R}{b} = \frac{110}{40} = 2.75 \not> \frac{W_L}{a} = \frac{60}{20} = 3.0$

Just to the left of C :

$W_L = 10 + 30 + 20 + 20 = 80$ k

$W_R = 110 - 20 = 90$ k .

$\frac{W_R}{b} = \frac{90}{40} = 2.25 < \frac{W_L}{a} = \frac{80}{20} = 4.0$ ✓

wheel ④ does not satisfy the criterion.

(H4)

Try wheel (5) placed at:

just to the right of C:

$w_L = 20 + 20 = 40$ k.

$w_R = 30 + 20 + 20 + 20 + 10 = 100$ k.

$\frac{w_R}{b} = \frac{100}{40} = 2.5 > \frac{w_L}{a} = \frac{40}{20} = 2.0$ ✓

just to the left of C:

$w_L = 40 + 30 = 70$ k.

$w_R = 100 - 30 = 70$ k.

$\frac{w_R}{b} = \frac{70}{40} = 1.75 < \frac{w_L}{a} = \frac{70}{20} = 3.5$ ✓

∴ wheel (5) <u>satisfies</u> the criterion.

Try wheel (6) placed at:

just to the right of C:

$w_L = 20 + 30 = 50$ k.

$w_R = 20 + 20 + 20 + 10 = 70$ k.

$\frac{w_R}{b} = \frac{70}{40} = 1.75 \not> \frac{w_L}{a} = \frac{50}{20} = 2.5$

just to the left of C:

$w_L = 50 + 20 = 70$ k.

$w_R = 70 - 20 = 50$ k

$\frac{w_R}{b} = \frac{50}{40} = 1.25 < \frac{w_L}{a} = \frac{70}{20} = 3.5$ ✓

∴ wheel (6) does <u>not</u> satisfy the criterion.

Find the bending moment at C for the <u>two</u> cases:

<u>Case (1)</u>: wheel (3) is at section C:

$$M_c = \Sigma(P_R \cdot x)\frac{h}{b} + \Sigma(P_L \cdot x)\frac{h}{a}$$

$$= [20(5) + 20(15) + 30(25) + 20(35) + 20(40)]\left(\frac{40/3}{40}\right)$$

$$+ [10(5) + 30(13)]\left(\frac{40/3}{20}\right)$$

$$= 883.333 + 293.333$$

$$= \underline{\underline{+1176.67}} \text{ k-ft.}$$

(H5)

Case (2): wheel ⑤ is at section C:

$$M_c = \Sigma(P_R \cdot x)\,h/b + \Sigma(P_L \cdot x)\,h/a$$

$$= [10(0) + 20(10) + 20(20) + 20(30) + 30(40)]\left(\frac{40/3}{40}\right)$$

$$+ [20(5) + 20(10)]\left(\frac{40/3}{20}\right)$$

$$= 800 + 200$$

$$= \underline{+1000} \text{ k-ft.}$$

$\therefore M_{c(max.)} = +1176.67$ k-ft when wheel ③ is at section C.

---

④ ⓐ $\bar{x} = \dfrac{16(6) + 20(12) + 4(16)}{10 + 16 + 20 + 4}$

$= 8'$.

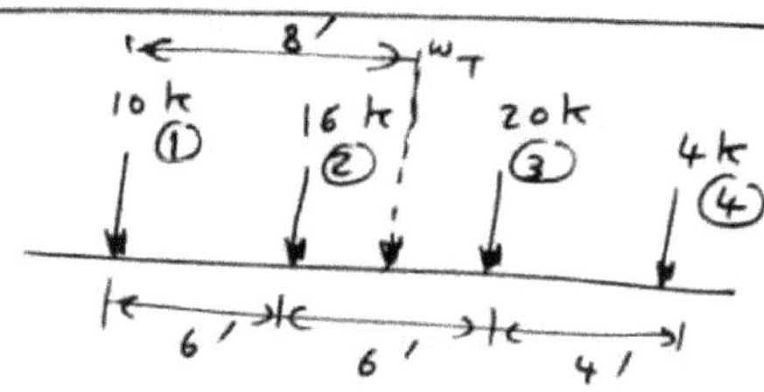

Place wheel ② at ℄:

just to the right of the section:

$w_L = 10$ k

$w_R = 16 + 20 + 4 = 40$ k.

$\frac{w_R}{b} = \frac{40}{10} = 4.0 > \frac{w_L}{a} = \frac{10}{10} = 1.0$ ✓

just to the left of the section:

$w_L = 10 + 16 = 26$ k.

$w_R = 20 + 4 = 24$ k.

$\frac{w_R}{b} = \frac{24}{10} = 2.4 < \frac{w_L}{a} = \frac{26}{10} = 2.6$

∴ wheel ② satisfies the criterion.

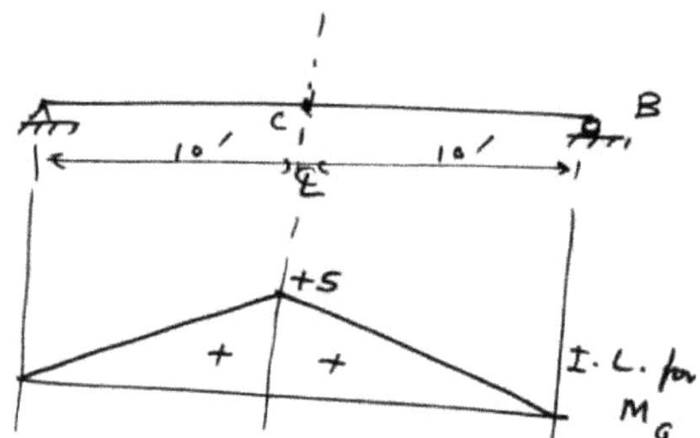

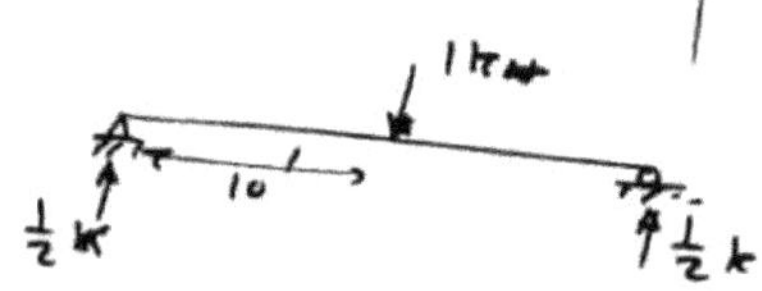

$M_℄ = \frac{1}{2}(10) = 5$ k-ft.

Place wheel ③ at ℄:

just to the right of the section:

$w_L = 16$ k

$w_R = 20 + 4 = 24$ k.

$\frac{w_R}{b} = \frac{24}{10} = 2.4 > \frac{w_L}{a} = \frac{16}{10} = 1.6$

just to the left of the section:

$w_L = 16 + 20 = 36$ k

$w_R = 4$ k

(46)

$\frac{w_R}{b} = \frac{4}{10} = 0.4 < \frac{w_L}{a} = \frac{36}{10} = 3.6$.

$\therefore$ Wheel ③ satisfies the criterion.

Find the bending moment at ℄ for the two cases:

Case (1): wheel ② is at ℄:

$$M_{℄} = \Sigma(P_R \cdot x)\frac{h}{b} + \Sigma(P_L \cdot x)\frac{h}{a}$$

$$= [4(0) + 20(4) + 16(10)]\left(\frac{5}{10}\right) + [10(4)]\left(\frac{5}{10}\right)$$

$$= 140 \text{ k-ft}.$$

Case (2): wheel ③ is at ℄:

$$M_{℄} = \Sigma(P_R \cdot x)\frac{h}{b} + \Sigma(P_L \cdot x)\frac{h}{a}$$

$$= [4(6) + 20(10)]\left(\frac{5}{10}\right) + [16(4)]\left(\frac{5}{10}\right)$$

$$= 144 \text{ k-ft}.$$

$\therefore$ $M_{℄(max.)} = +144$ k-ft when wheel ③ is at the center-line

ⓑ It is obvious that case (2) will produce the absolute maximum bending moment (with only three wheels on the beam).

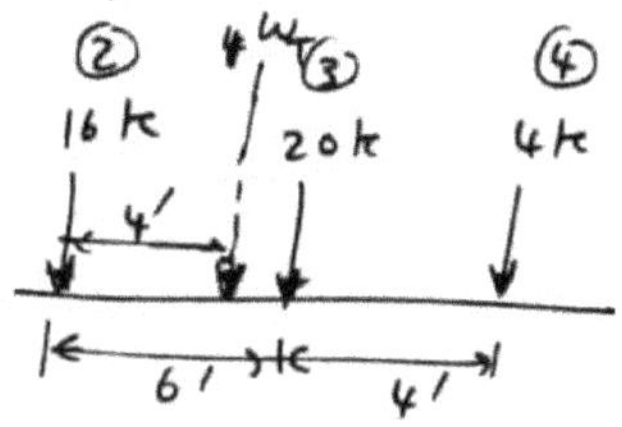

$$\therefore \bar{x}_1 = \frac{20(6) + 4(10)}{16 + 20 + 4}$$

$$= 4'.$$

For the absolute maximum bending moment, place the wheels on the span as shown in the figure.

$$x = \frac{L}{2} + \frac{a}{2}$$

$$= 10 + \frac{6-4}{2}$$

$$= 11'.$$

Check if the criterion is satisfied:

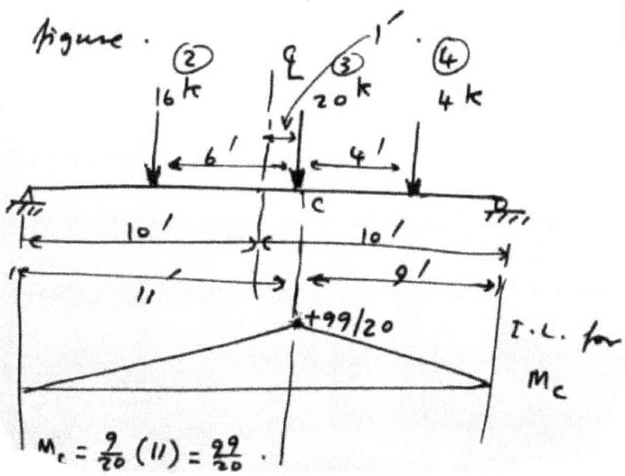

(H7)

the wheel just to the right of C :

$w_L = 16\ k$

$w_R = 20 + 4 = 24\ k$.

$$\frac{w_R}{b} = \frac{24}{9} = 2.67 > \frac{w_L}{a} = \frac{16}{11} = 1.45 \quad ✓$$

the wheel just to the left of C :

$w_L = 16 + 20 = 36\ k$.

$w_R = 4\ k$.

$$\frac{w_R}{b} = \frac{4}{9} = 0.44 < \frac{w_L}{a} = \frac{36}{11} = 3.27 \quad ✓$$

∴ wheel ③ satisfies the criterion.

Absolute Maximum $M = \Sigma(P_R \cdot x)\frac{h}{b} + \Sigma(P_L \cdot x)\frac{h}{a}$

$$= [4(5) + 20(9)]\left(\frac{99/20}{9}\right) + [16(5)]\left(\frac{99/20}{11}\right)$$

$$= 110 + 36$$

$$= \underline{\underline{+146}}\ \text{k-ft}.$$

(H1)

Homework 10

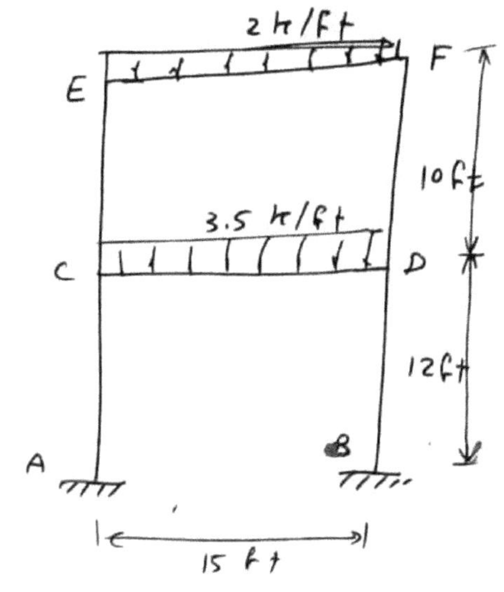

① Determine (approximately) the support reactions at A and B due to the uniform vertical load acting on the frame.

Solution:

the frame is indeterminate to the 6th degree.

We introduce four hinges in the two beams at 0.1L from each support. and assume zero axial force in each beams.

$0.1L = 0.1(15) = 1.5$ ft.

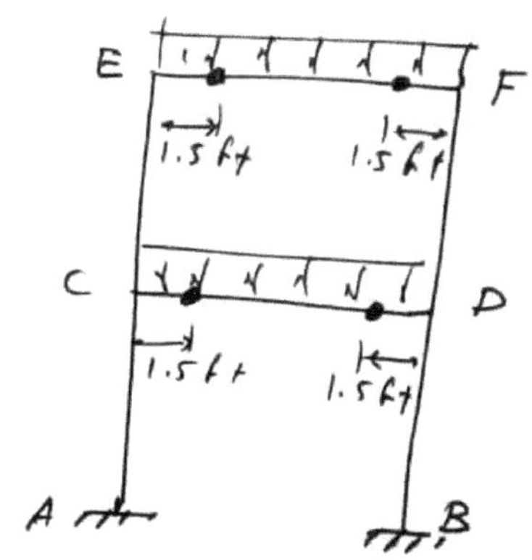

Top beam between hinges:

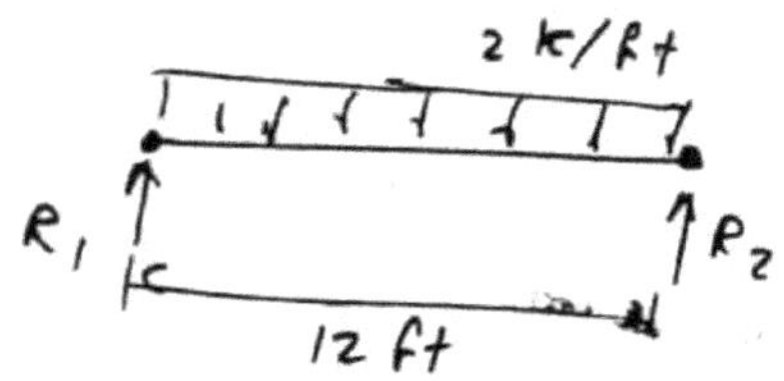

$$R_1 = R_2 = \frac{2(12)}{2} = 12 \text{ k}.$$

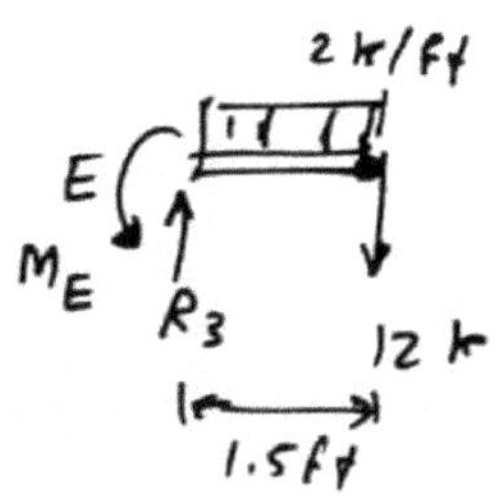

$$+\circlearrowright \Sigma M_E = 0: M_E - 2(1.5)\left(\frac{1.5}{2}\right) - 12(1.5) = 0$$

$$M_E = 20.25 \text{ k-ft}.$$

$$R_3 = 12 + 2(1.5) = 15 \text{ k}.$$

Joint E:

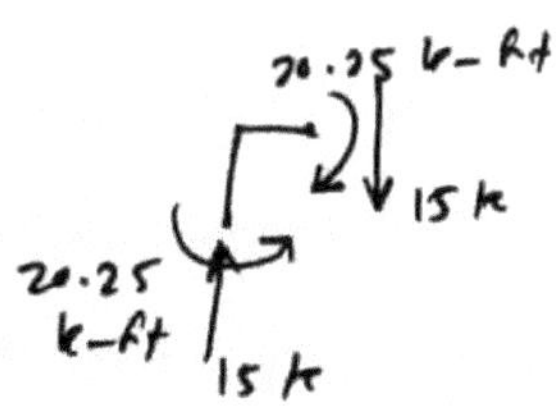

(H2)

Middle Beam between hinges:

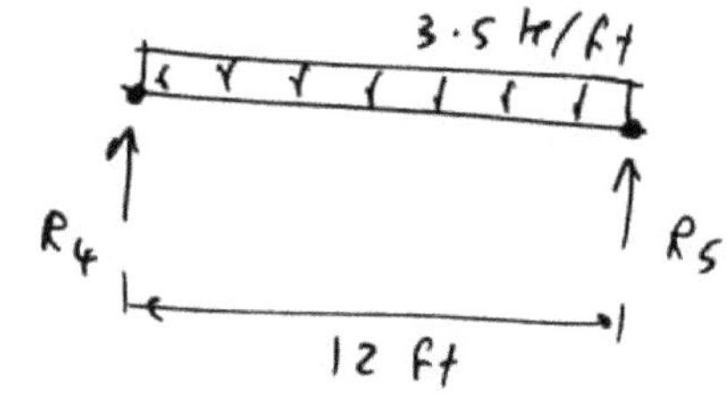

$$R_4 = R_5 = \frac{3.5(12)}{2} = 21 \text{ k}.$$

$$R_6 = 21 + 3.5(1.5) = 26.25 \text{ k}.$$

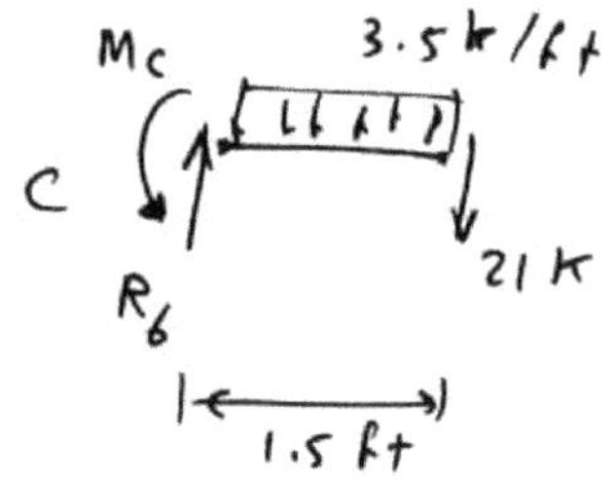

$$+\circlearrowright \Sigma M_c = 0: \; M_c - 3.5(1.5)\left(\frac{1.5}{2}\right) - 21(1.5) = 0$$

$$M_c = 35.44 \text{ k-ft}.$$

Joint C:

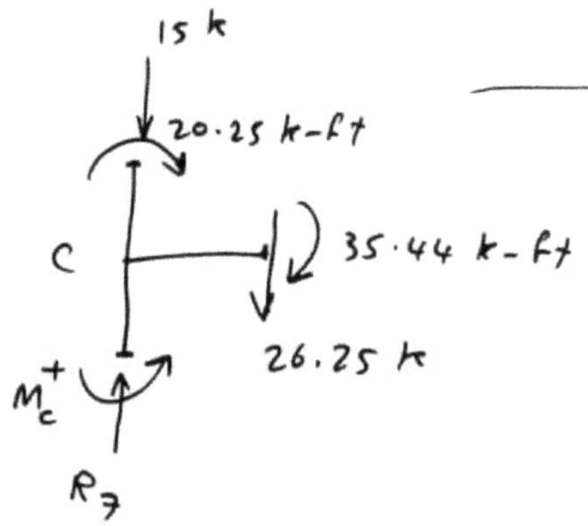

$$R_7 = 15 + 26.23 = 41.25$$

$$M_c^+ = 35.44 + 20.23 = 55.69 \text{ k-ft}.$$

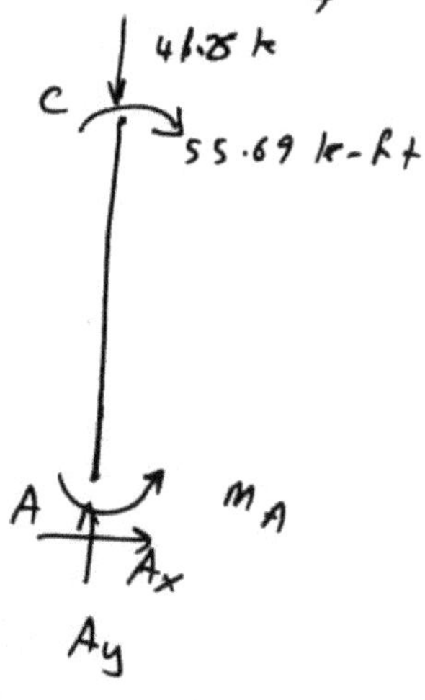

$A_x = 0$

$A_y = 41.25$ k.

$M_A = 55.69$ k-ft.

By symmetry, $B_x = A_x = 0$.

$B_y = A_y = 41.25$ k.

$M_B = M_A = 55.69$ k-ft.

(2) Using the portal method of approximate analysis, determine the reactions at the base of the columns of the frame shown in the figure. Show the reactions on a sketch of the frame. (H3)

Solution:

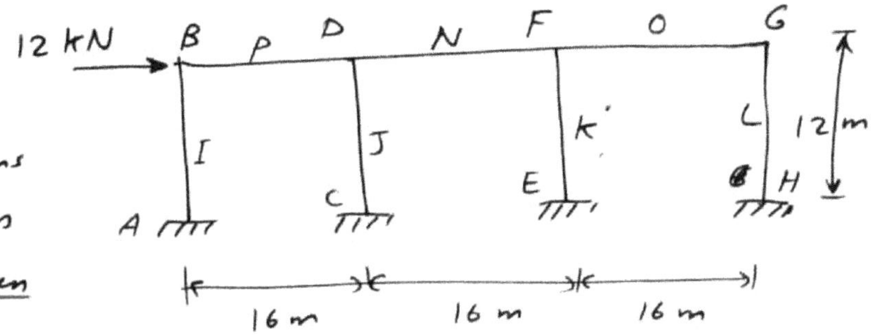

First, place hinges at the centers of the beams and columns, i.e. hinges are placed at the seven locations

I, J, K, L, P, N, O.

The frame is indeterminate to the 9th degree.

∴ We use assumption (3) of the portal method, at the floor level I, J, K, L:

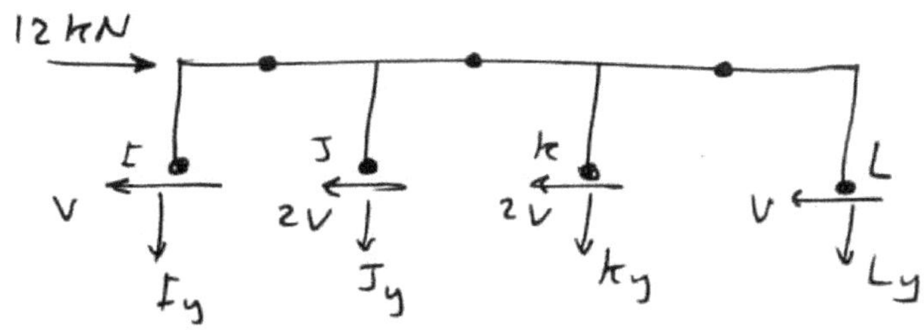

$\overset{+}{\rightarrow} \Sigma F_x = 0:$

$12 - V - 2V - 2V - V = 0$

$V = 2$ kN.

Section IP:

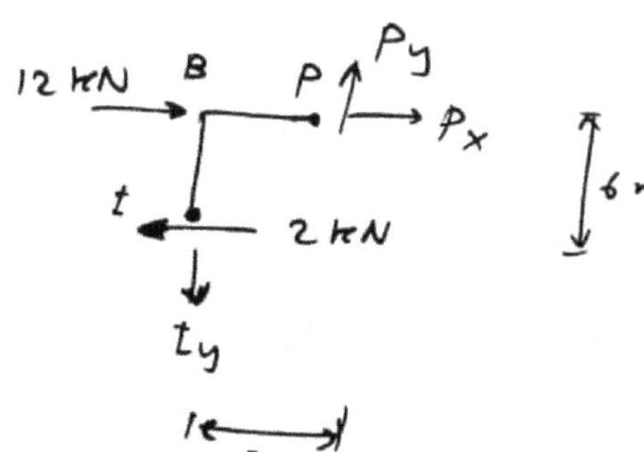

$\overset{+}{\rightarrow} \Sigma F_x = 0:$

$P_x + 12 - 2 = 0$

$P_x = -10$ kN.

$+\circlearrowleft \Sigma M_I = 0:$

$-(-10)(6) - 12(6) + P_y(8) = 0$

$P_y = 1.5$ kN.

$+\uparrow \Sigma F_y = 0:$

$-I_y + 1.5 = 0$

$I_y = 1.5$ kN.

(H4)

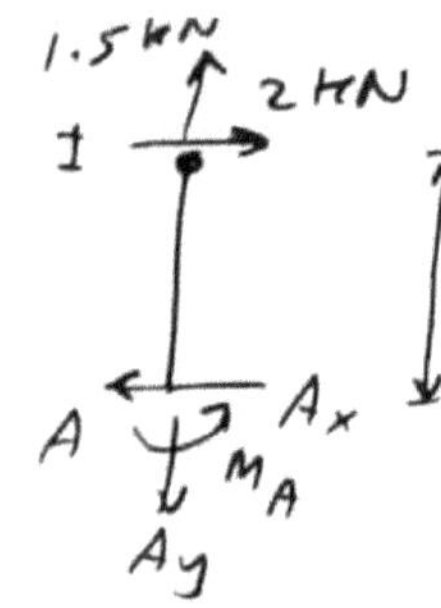

Section IA:

$\xrightarrow{+} \Sigma F_x = 0: \quad \boxed{A_x = 2 \text{ kN}}$

$+\uparrow \Sigma F_y = 0: \quad \boxed{A_y = 1.5 \text{ kN}}$

$+\circlearrowleft \Sigma M_A = 0: \quad M_A - 2(6) = 0$

$\boxed{M_A = 12 \text{ kN}\cdot\text{m}}$

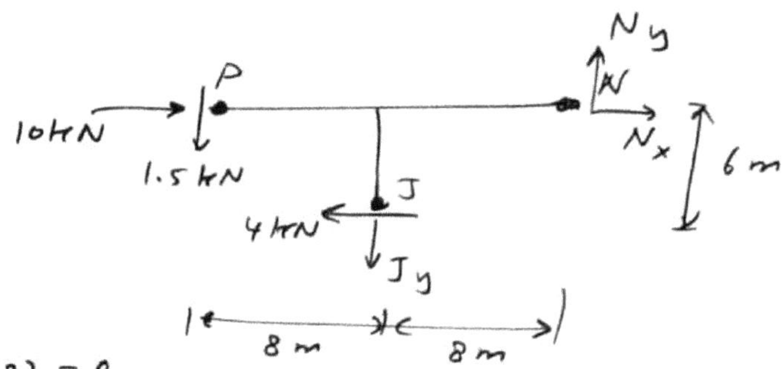

Section PJN:

$\xrightarrow{+} \Sigma F_x = 0:$

$10 - 4 + N_x = 0$

$N_x = -6 \text{ kN}.$

$+\circlearrowleft \Sigma M_J = 0:$

$-(-6)(6) + N_y(8) - 10(6) + 1.5(8) = 0$

$N_y = 1.5 \text{ kN}.$

$+\uparrow \Sigma F_y = 0: \quad -1.5 - J_y + 1.5 = 0 \Rightarrow J_y = 0.$

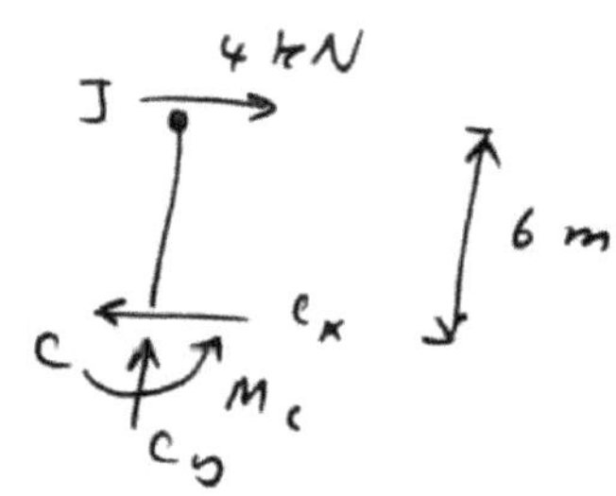

Section JC:

$\boxed{C_x = 4 \text{ kN}}$

$\boxed{C_y = 0}$

$+\circlearrowleft \Sigma M_C = 0: \quad M_C - 4(6) = 0$

$\boxed{M_C = 24 \text{ kN}\cdot\text{m}}$

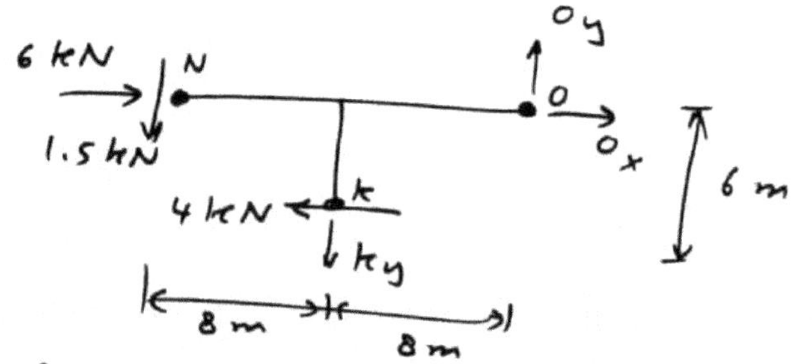

Section NKO:

$\xrightarrow{+} \Sigma F_x = 0:$

$6 - 4 + O_x = 0$

$O_x = -2 \text{ kN}.$

$+\circlearrowleft \Sigma M_K = 0:$

$-(-2)(6) + O_y(8) - 6(6) + 1.5(8) = 0$

$O_y = 1.5 \text{ kN}.$

$+\uparrow \Sigma F_y = 0:$

$-1.5 + K_y + 1.5 = 0$

$K_y = 0.$

(H5)

Section KE :

$\boxed{E_x = 4 \text{ kN}}$

$\boxed{E_y = 0}$

$M_E = 4(6) = 24 \text{ kN}\cdot\text{m}$

$\boxed{M_E = 24 \text{ kN}\cdot\text{m}}$

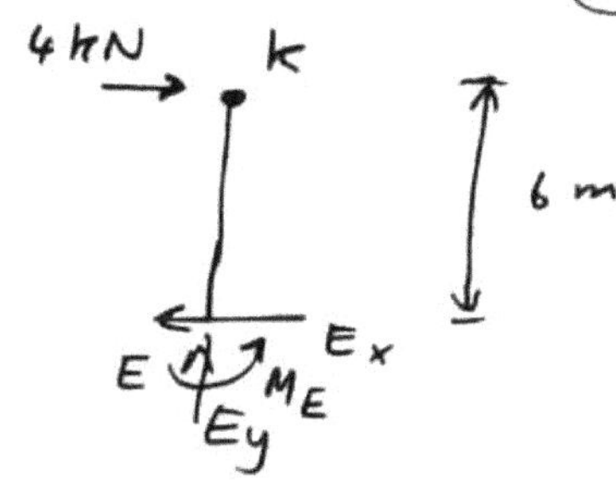

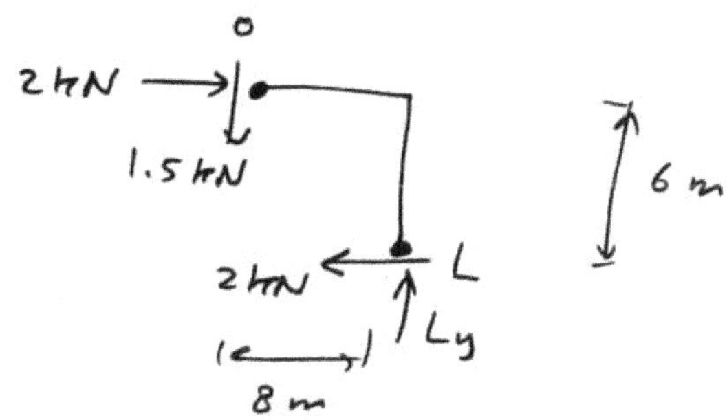

Section OL :

$+\uparrow \Sigma F_y = 0: \quad L_y = 1.5 \text{ kN}.$

Check : $+\circlearrowleft \Sigma M_L = 0$:

$1.5(8) - 2(6) = 0$

$0 = 0$ ✓

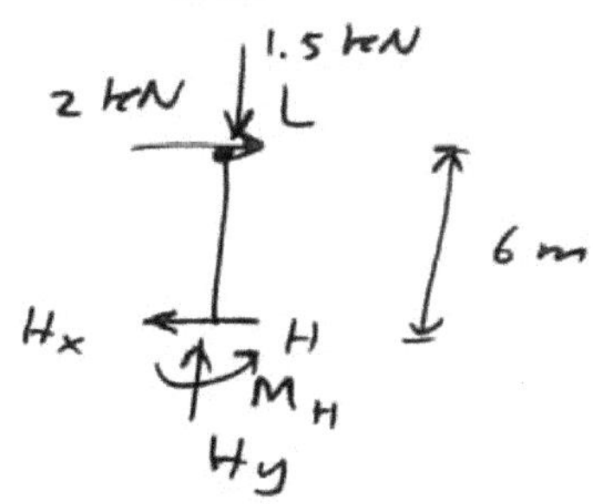

Section LH :

$\boxed{H_x = 2 \text{ kN}}$

$\boxed{H_y = 1.5 \text{ kN}}$

$M_H = 2(6) = 12 \text{ kN}\cdot\text{m}$

$\boxed{M_H = 12 \text{ kN}\cdot\text{m}}$

the final results are shown below:

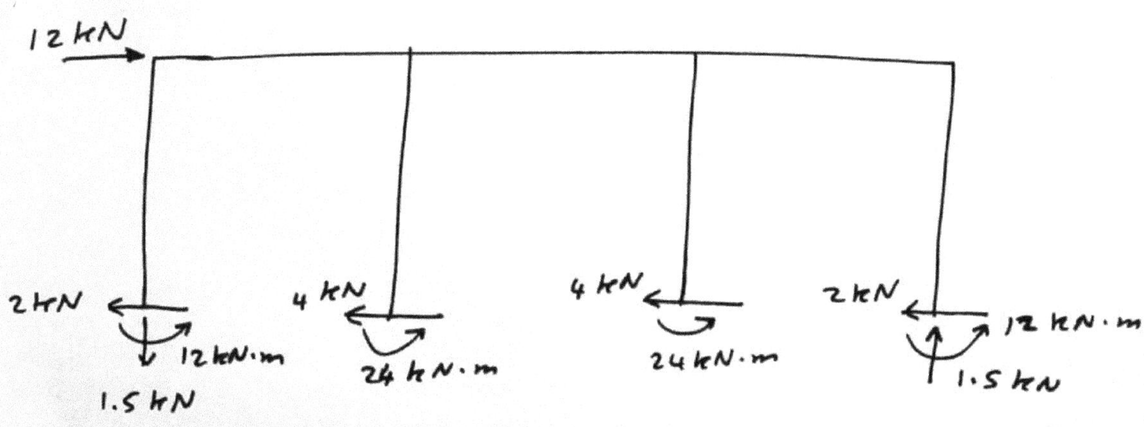

(3) Using the cantilever method of approximate analysis, determine the reactions at the base of the columns of the frame shown in the figure. Show the reactions on a sketch of the frame. (H6)

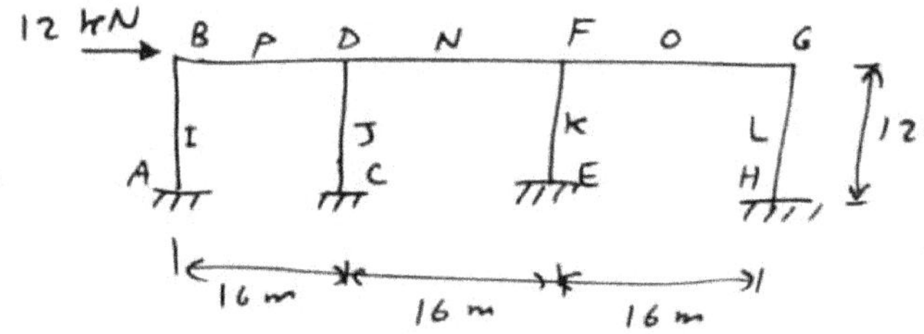

Solution :

First, locate the centroid of the columns' cross-sectional areas.

Let A ≡ area of each column.

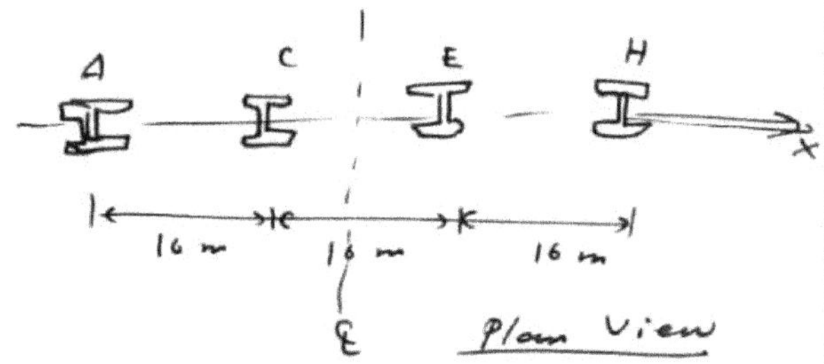

$$\bar{x} = \frac{\Sigma \tilde{x} A}{\Sigma A}$$

$$= \frac{0(A)+16(A)+32(A)+48(A)}{A+A+A+A}$$

$$= \frac{0+16+32+48}{4}$$

$$= 24 \text{ m}.$$

Second, place hinges at the center of each beam and column.

Assume that the axial force in each column is proportional to its distance from the centroid.

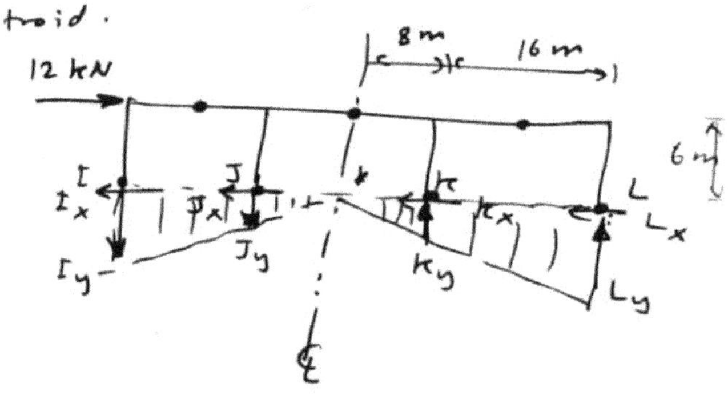

From similar triangles,

$$\frac{L_y}{24} = \frac{I_y}{24} \Rightarrow L_y = I_y$$

$$\frac{K_y}{8} = \frac{I_y}{24} \Rightarrow K_y = \frac{1}{3} I_y$$

$$\frac{J_y}{8} = \frac{I_y}{24} \Rightarrow J_y = \frac{1}{3} I_y$$

$$\circlearrowleft + \Sigma M_I = 0: \quad L_y(48) + K_y(32) - J_y(16) - 12(6) = 0$$

$$\therefore \quad I_y(48) + \frac{1}{3} I_y(32) - \frac{1}{3} I_y(16) - 72 = 0$$

$$\left(48 + \frac{32}{3} - \frac{16}{3}\right) I_y = 72 \Longrightarrow \underline{I_y = 1.35 \text{ kN}}.$$

$$\therefore \quad L_y = I_y = 1.35 \text{ kN}.$$

$$K_y = J_y = \frac{1}{3}(1.35) = 0.45 \text{ kN}.$$

(H7)

Section IP :

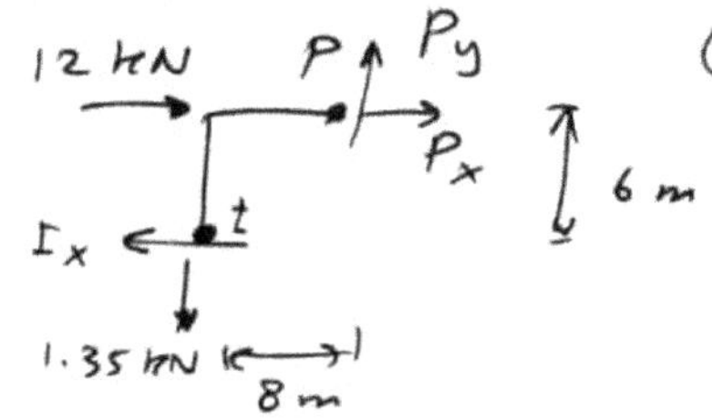

$+\uparrow \Sigma F_y = 0:$

$P_y = 1.35$ kN

$+\circlearrowleft \Sigma M_I = 0:$

$-P_x(6) - 12(6) + 1.35(8) = 0$

$P_x = -10.2$ kN.

$\xrightarrow{+} \Sigma F_x = 0: \quad -I_x + 12 + (-10.2) = 0$

$I_x = 1.8$ kN.

Section IA :

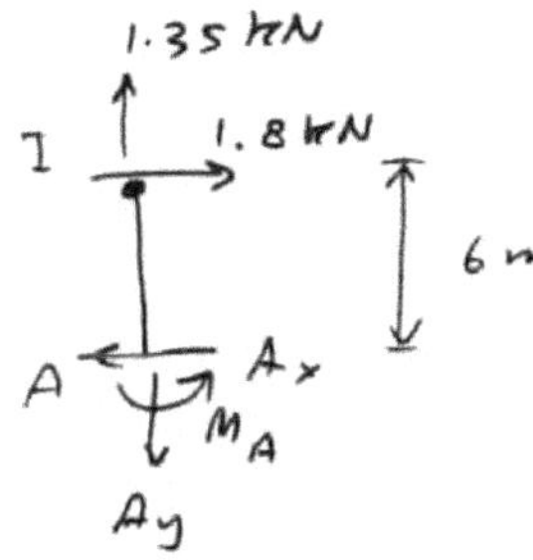

$\boxed{A_x = 1.8 \text{ kN}}$

$\boxed{A_y = 1.35 \text{ kN}}$

$M_A = 1.8(6) = 10.8$ kN·m

$\boxed{M_A = 10.8 \text{ kN·m}}$

Section PJN :

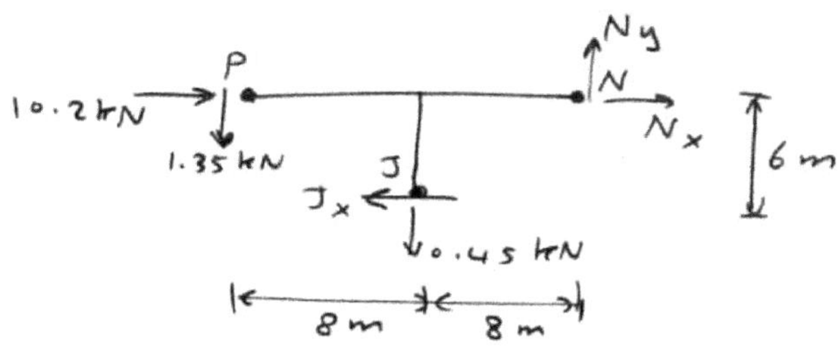

$+\uparrow \Sigma F_y = 0:$

$-0.45 - 1.35 + N_y = 0$

$N_y = 1.8$ kN.

$+\circlearrowleft \Sigma M_J = 0:$

$-N_x(6) - 10.2(6) + (1.8)(8) + 1.35(8) = 0$

$N_x = -6$ kN.

$\xrightarrow{+} \Sigma F_x = 0: \quad -J_x + (-6) + 10.2 = 0$

$J_x = 4.2$ kN.

Section JC :

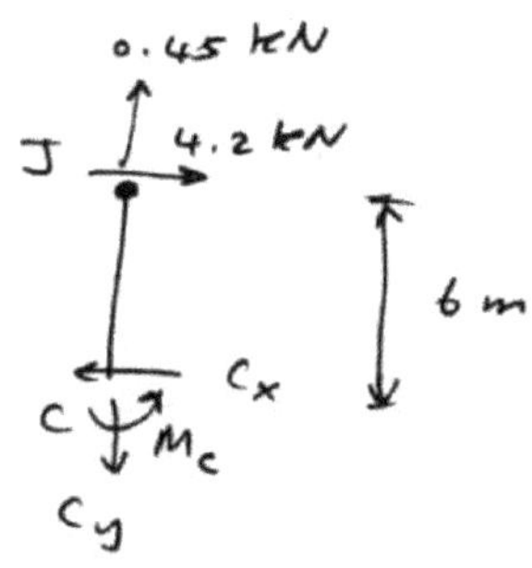

$\boxed{C_x = 4.2 \text{ kN}}$

$\boxed{C_y = 0.45 \text{ kN}}$

$M_c = 4.2(6) = 25.2$ kN·m

$\boxed{M_c = 25.2 \text{ kN·m}}$

Section NKO :

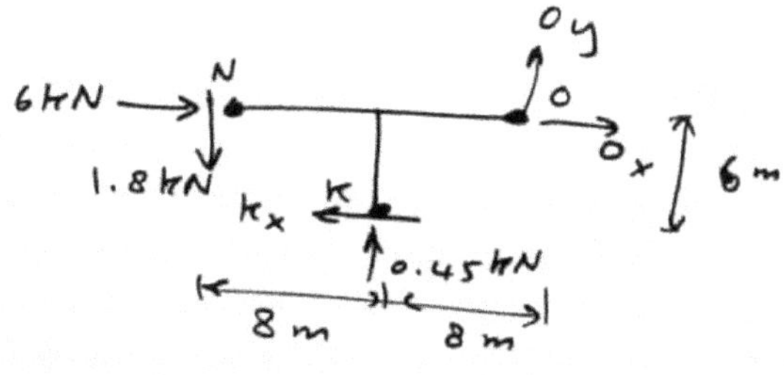

$+\uparrow \Sigma F_y = 0:$

$-1.8 + 0.45 + O_y = 0$

$O_y = 1.35$ kN

(H8)

$+\circlearrowleft \; \Sigma M_K = 0:$

$$-O_x(6) - 6(6) + (1.35)(8) + 1.8(8) = 0$$

$$O_x = -1.8 \text{ kN}.$$

$\xrightarrow{+} \Sigma F_x = 0: \quad 6 - K_x + (-1.8) = 0$

$$K_x = 4.2 \text{ kN}.$$

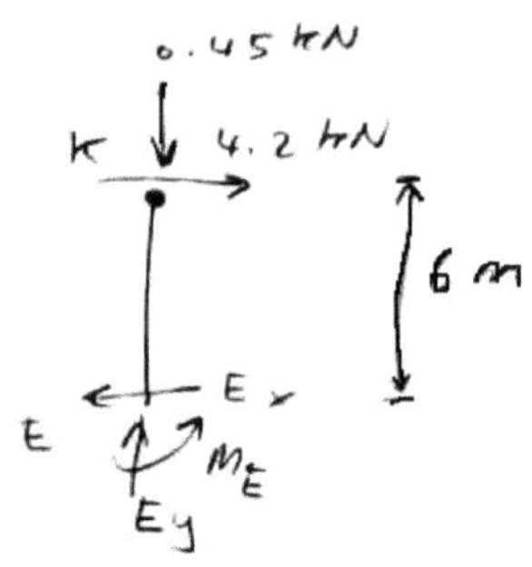

Section KE :

$$\boxed{E_x = 4.2 \text{ kN}}$$

$$\boxed{E_y = 0.45 \text{ kN}}$$

$$M_E = 4.2(6) = 25.2 \text{ kN}\cdot\text{m}$$

$$\boxed{M_E = 25.2 \text{ kN}\cdot\text{m}}$$

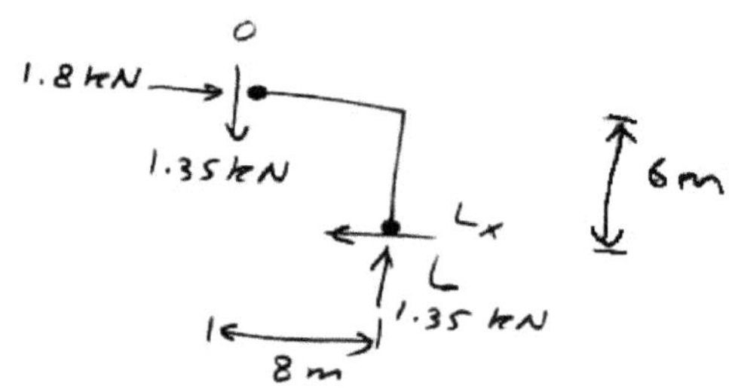

Section OL :

$\xrightarrow{+} \Sigma F_x = 0:$

$$L_x = 1.8 \text{ kN}.$$

Check: $+\circlearrowleft \Sigma M_L = 0:$

$$1.35(8) - 1.8(6) = 0$$

$$0 = 0 \;\checkmark$$

The results are shown below :

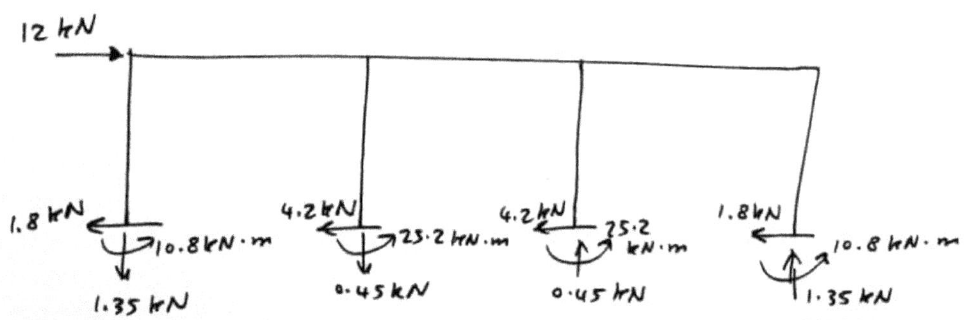

Section LH :

$$\boxed{H_x = 1.8 \text{ kN}}$$

$$\boxed{H_y = 1.35 \text{ kN}}$$

$$M_H = 1.8(6) = 10.8 \text{ kN}\cdot\text{m}$$

$$\boxed{M_H = 10.8 \text{ kN}\cdot\text{m}}$$

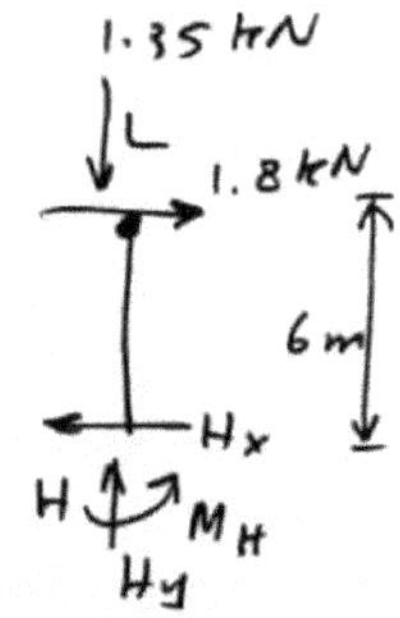

(H9)

The following table shows a comparison of the results with a computer solution.

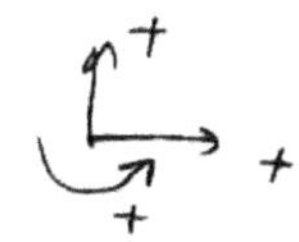

Reaction	Portal Method	Cantilever Method	Computer Solution
A_x	−2.0	−1.8	−2.63
A_y	−1.5	−1.35	−1.43
M_A	+12.0	+10.8	+19.1
C_x	−4.0	−4.2	−3.37
C_y	0	−0.45	+0.41
M_C	+24.0	+25.2	+22.0
E_x	−4.0	−4.2	−3.37
E_y	0	+0.45	−0.41
M_E	+24.0	+25.2	+22.0
H_x	−2.0	−1.8	−2.63
H_y	+1.5	+1.35	+1.43
M_H	+12.0	+10.8	+19.1

(H1)

Homework 11

Problem 11.2 :

Solve for the reactions and draw the bending-moment diagram.

Solution : I = constant.

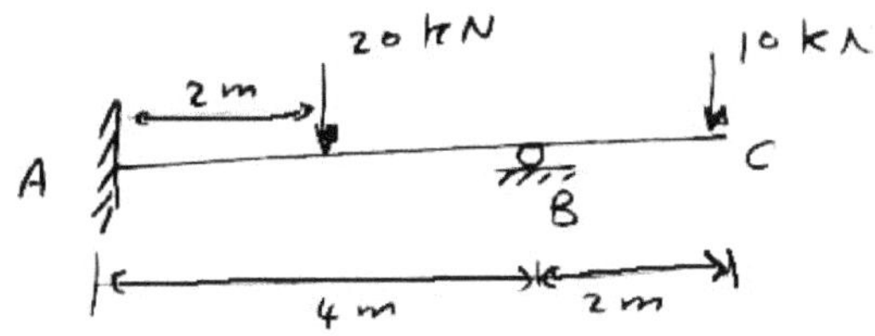

By superposition :

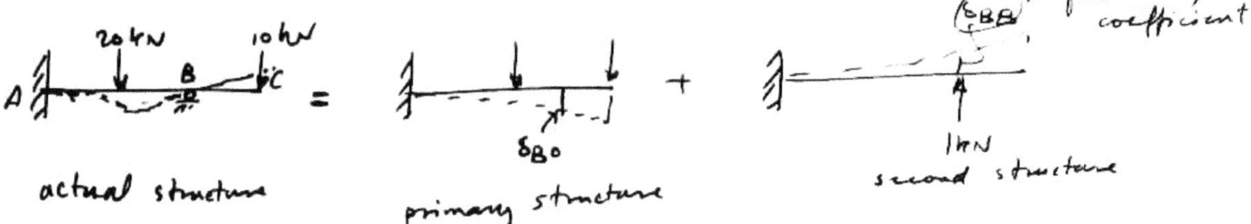

(1) primary structure:

use the moment-area theorems:

$$M_A = 10(6) + 20(2) = 100 \text{ kN·m}$$

$$\delta_{B0} = \frac{20}{EI}(2)(1) + \frac{1}{2}\left(\frac{20}{EI}\right)(2)\left(\frac{2}{3}\right)(2) + \frac{40}{EI}(2)(3) + \frac{1}{2}\left(\frac{60}{EI}\right)(2)\left(2 + \frac{2}{3}(2)\right)$$

$$= \frac{506.67}{EI} \text{ (down)}.$$

take $\delta_{B0} = -\dfrac{506.67}{EI}$.

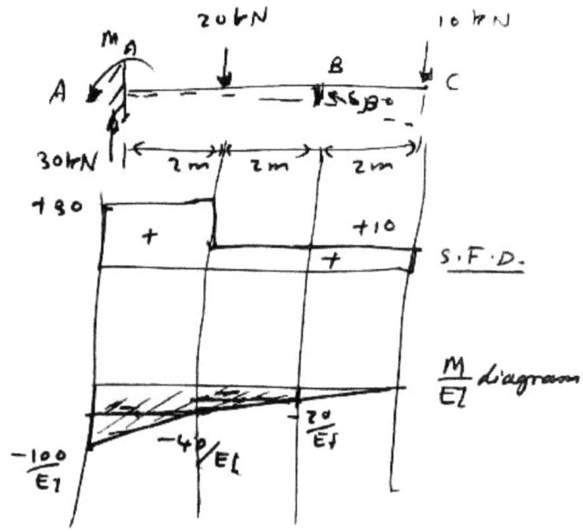

(2) second structure:

$$\delta_{BB} = \frac{1}{2}\left(\frac{4}{EI}\right)(4)\left(\frac{2}{3}\right)(4) = \frac{21.33}{EI}.$$

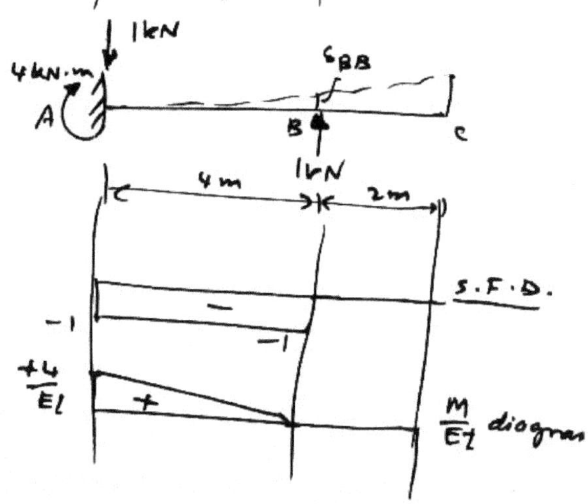

(2) second structure.

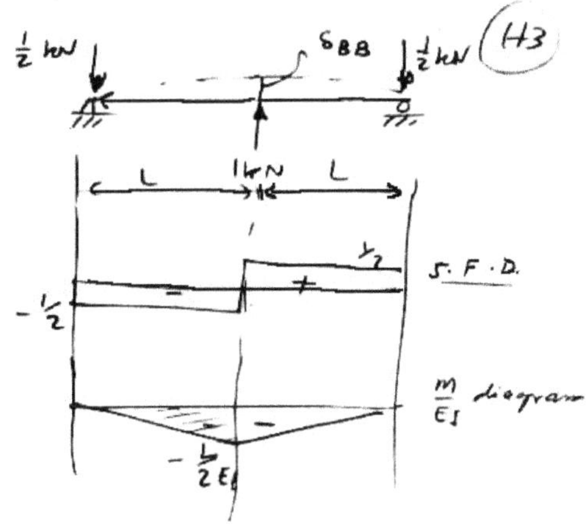

$$\delta_{BB} = \frac{1}{2}\left(\frac{L}{2EI}\right)(L)\left(\frac{2}{3}L\right)$$

$$= \frac{L^3}{6EI} \quad \text{(up)}$$

(3) equation of consistent deformation:

$$\delta_B = \delta_{B0} + R_B\delta_{BB} = 0$$

$$-\frac{5wL^4}{24EI} + R_B\left(\frac{L^3}{6EI}\right) = 0$$

$$R_B = \frac{5wL^4}{24EI}\cdot\frac{6EI}{L^3} = \frac{5wL}{4}\ \uparrow.$$

(4) equations of equilibrium:

$\circlearrowleft\ \Sigma M_A = 0$:

$$\frac{5wL}{4}(L) + R_C(2L) - w(2L)(L) = 0$$

$$2LR_C = 2wL^2 - \frac{5wL^2}{4}$$

$$2LR_C = \frac{3wL^2}{4}$$

$$R_C = \frac{3wL}{8}\ \uparrow.$$

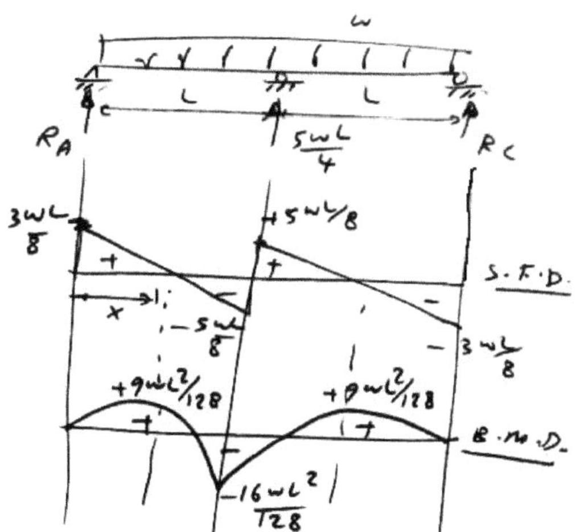

$\uparrow\ \Sigma F_y = 0$: $R_A + \frac{5wL}{4} + \frac{3wL}{8} - w(2L) = 0$

$$R_A = 2wL - \frac{5wL}{4} - \frac{3wL}{8}$$

$$\Rightarrow R_A = \frac{3wL}{8}\ \uparrow. \quad \text{(by symmetry)}.$$

$$\frac{3}{8} - 1 = -\frac{5}{8}$$

$$-\frac{5}{8} + \frac{5}{4} = -\frac{5}{8} + \frac{10}{8} = \frac{5}{8}$$

$$\frac{5}{8} - 1 = -\frac{3}{8}$$

* Determine x (point of zero shear):

$$\frac{x}{L-x} = \frac{3wL/8}{5wL/8} \Rightarrow \frac{x}{L-x} = \frac{3}{5} \Rightarrow 5x = 3L - 3x$$

$$8x = 3L \Rightarrow x = \frac{3}{8}L.$$

$$M_{max.} = \frac{1}{2}\left(\frac{3wL}{8}\right)\left(\frac{3}{8}L\right) = \frac{9wL^2}{128}$$

$$M_{\substack{\text{inner}\\\text{support}}} = \frac{9wL^2}{128} - \frac{1}{2}\left(\frac{5wL}{8}\right)\left(\frac{5L}{8}\right) = \frac{9wL^2}{128} - \frac{25wL^2}{128} = -\frac{16wL^2}{128}.$$

Problem 11.10 :

(H4)

Solve for the reactions and draw the bending-moment diagram.

I = constant.

Solution :

Remove support C.

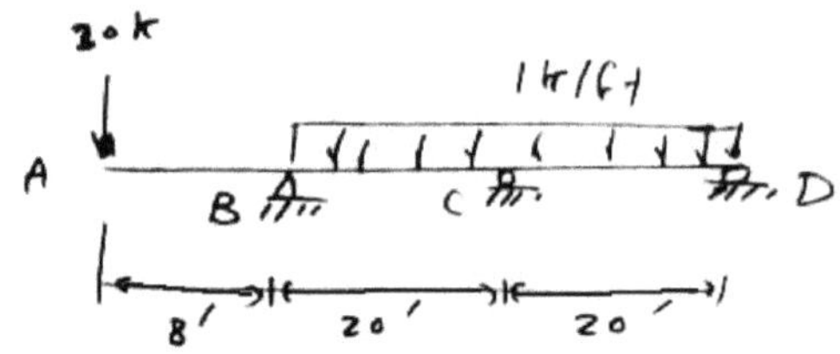

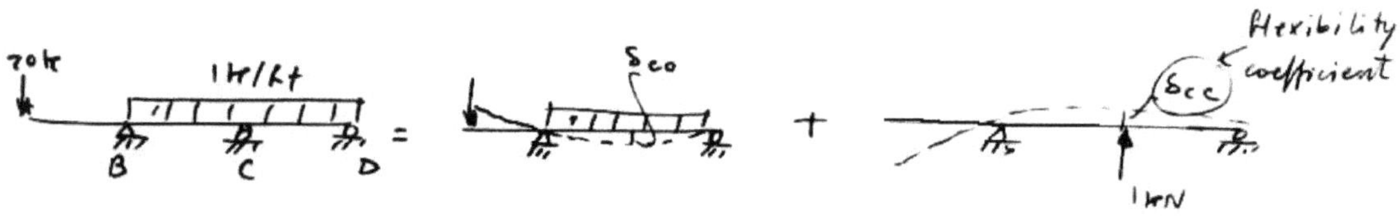

actual structure — primary structure — second structure

(1) primary structure :

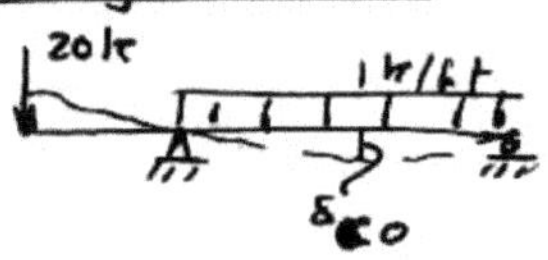

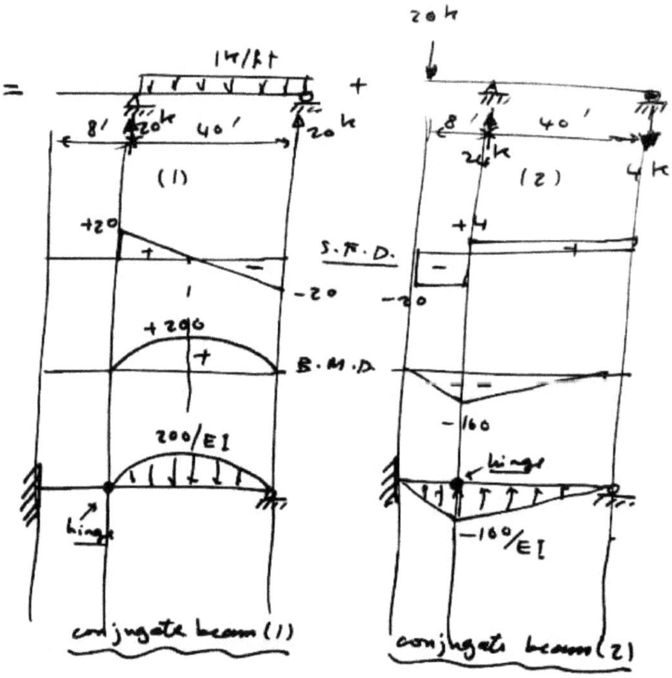

Conjugate Beam (1) :

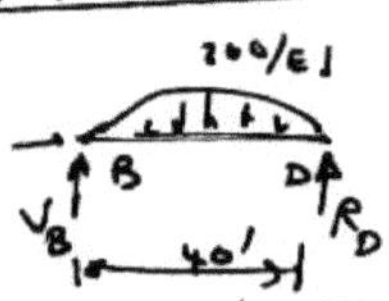

$$V_B = R_D = \frac{1}{2}(2)\left(\frac{2}{3}\right)\left(\frac{200}{EI}\right)(20)$$

$$= \frac{2666.67}{EI}.$$

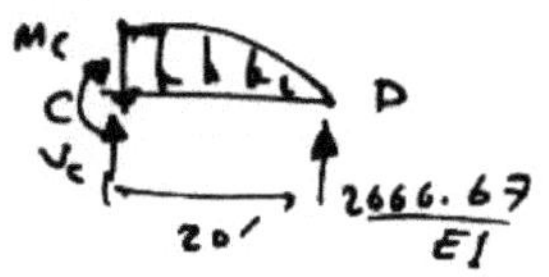

$$M_c = \frac{2666.67}{EI}(20) - \frac{2}{3}\left(\frac{200}{EI}\right)(20)\left(\frac{3}{8}\right)(20) = \frac{33,333.4}{EI}.$$

$$\therefore \ \delta_{co_1} = \frac{33,333.4}{EI} \ \checkmark \ \text{(down)}$$

Conjugate Beam (2) :

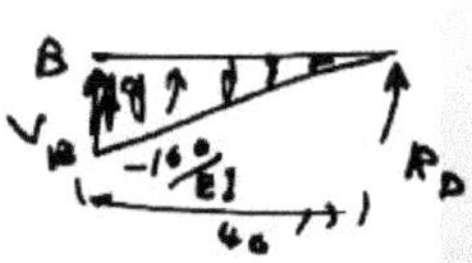

$$+\circlearrowleft \Sigma M_B = 0:$$

$$R_D(40) + \frac{1}{2}\left(\frac{160}{EI}\right)(40)\left(\frac{1}{3}\right)(40) = 0$$

$$\Rightarrow R_D = -\frac{1066.67}{EI}.$$

(H5)

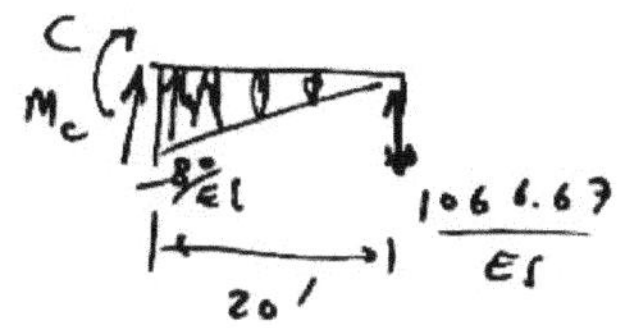

$$M_c = -\frac{1066.67}{EI}(20) + \frac{1}{2}\left(\frac{80}{EI}\right)(20)\left(\frac{1}{3}\right)(20)$$
$$= \frac{-16{,}000.07}{EI}$$

$$\therefore \delta_{co_2} = -\frac{16{,}000.07}{EI} \quad (\text{up})$$

$$\therefore \delta_{co} = \delta_{co_1} + \delta_{co_2} = \frac{33{,}333.4}{EI} - \frac{16{,}000.07}{EI} = \frac{17{,}333.33}{EI} \quad (\text{down})$$

(2) second structure:

Consider the conjugate beam:

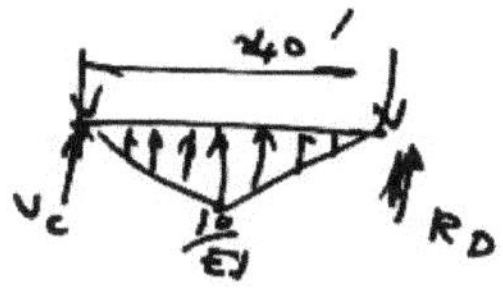

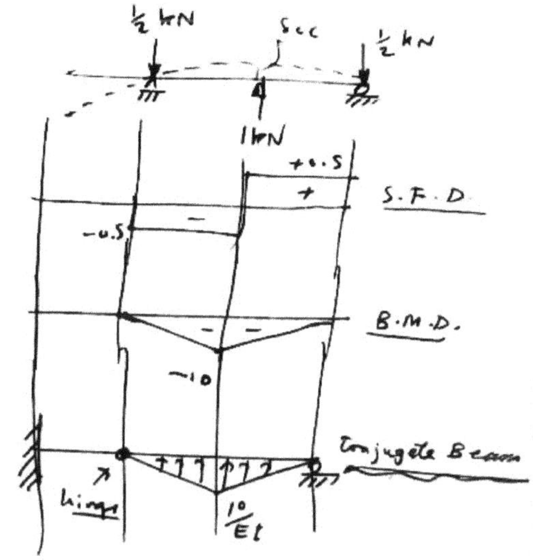

$$R_D = V_c = \frac{1}{2}\left(\frac{10}{EI}\right)(20) = \frac{100}{EI}$$

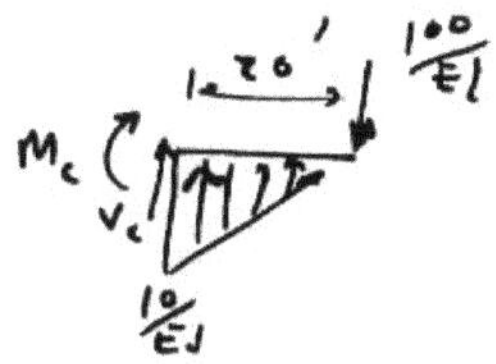

$$M_c = -\frac{100}{EI}(20) + \frac{1}{2}\left(\frac{10}{EI}\right)(20)\left(\frac{1}{3}\right)(20) = -\frac{1333.33}{EI}$$

$$\therefore \delta_{cc} = \frac{-1333.33}{EI} \quad (\text{up})$$

(3) equation of consistent deformations:

$$\delta_c = 0 = \delta_{co} + R_c\,\delta_{cc} = 0$$

$$\frac{17{,}333.33}{EI} + R_c\left(-\frac{1333.33}{EI}\right) = 0$$

$$R_c = 13\ \text{k} \uparrow$$

(4) equations of equilibrium:

$+\circlearrowright\ \Sigma M_B = 0:$

$$20(8) + 13(20) + R_D(40) - 1(40)(20) = 0$$

$$R_D = 9.5\ \text{k} \uparrow$$

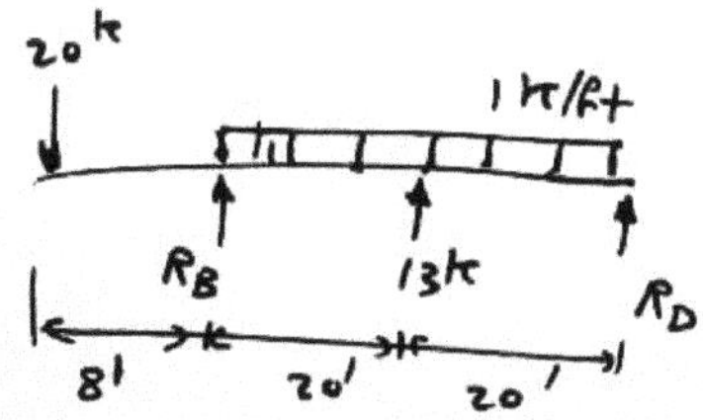

(H6)

$+\circlearrowleft \Sigma F_y = 0: \quad -20 + R_B + 13 + 9.5 - 40 = 0$

$R_B = 37.5 \text{ k} \uparrow.$

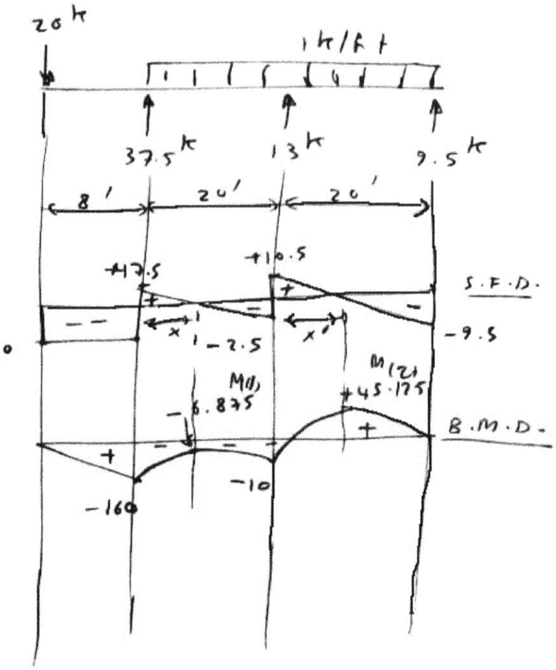

Determine x (point of zero shear):

$$\frac{x}{20-x} = \frac{17.5}{2.5}$$

$17.5(20) - 17.5x = 2.5x$

$20x = 17.5(20)$

$x = 17.5'$.

$$M_{(1)} = -160 + \frac{1}{2}(17.5)(17.5)$$

$= -6.875$

Determine x': (point of zero shear)

$$\frac{x}{20-x} = \frac{10.5}{9.5}$$

$10.5(20) - 10.5x = 9.5$

$20x = 10.5(20)$

$x = 10.5'$.

$M_{(2)} = -10 + \frac{1}{2}(10.5)(10.5) = 45.125.$

$V_{max.} = -20 \text{ k}.$

$M_{max.} = -160 \text{ k}$

Problem 11.12:

Solve for the reactions and draw the bending-moment diagram.

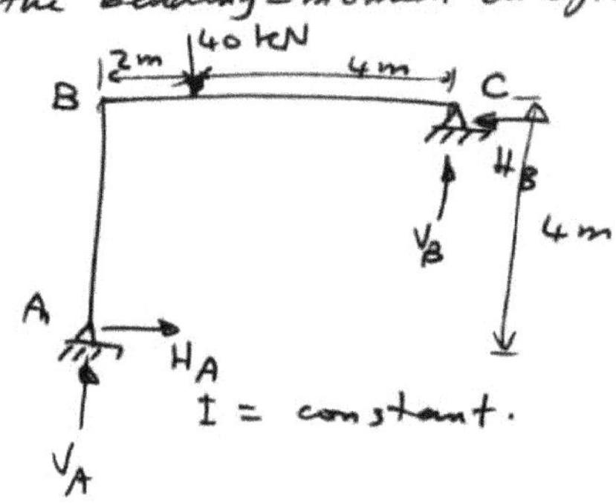

Solution:

Remove the horizontal reaction at A.

Consider H_A as the redundant.

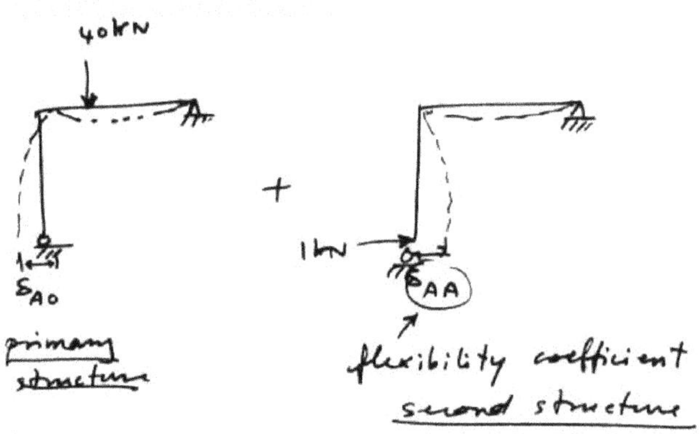

(H7)

(1) Primary Structure:

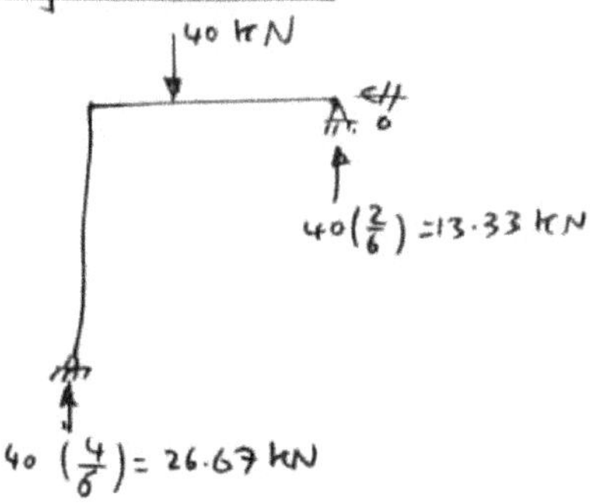

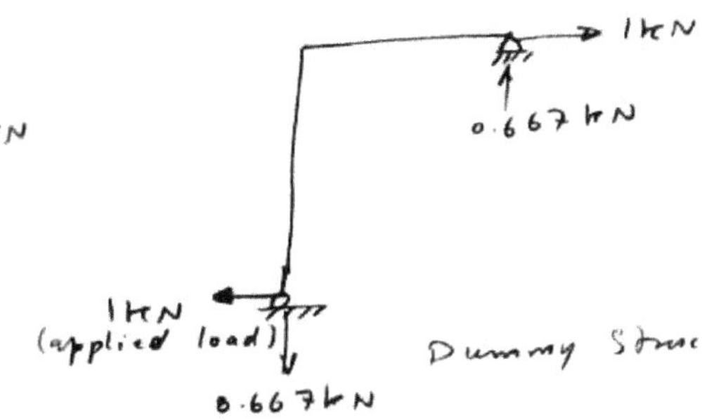

Dummy Structure

Member AB:

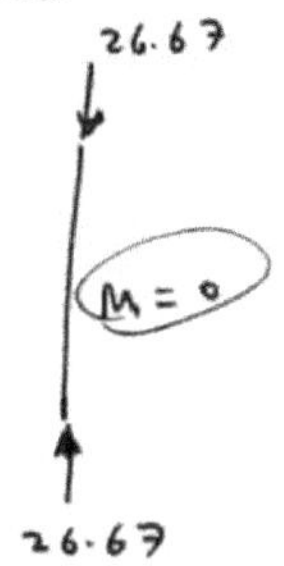

$$\Rightarrow \int \frac{Mm\,dx}{EI} = 0$$

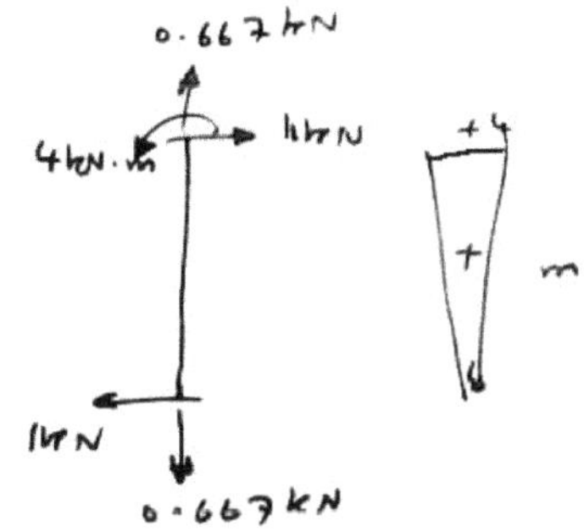

Dummy member

Member BC:

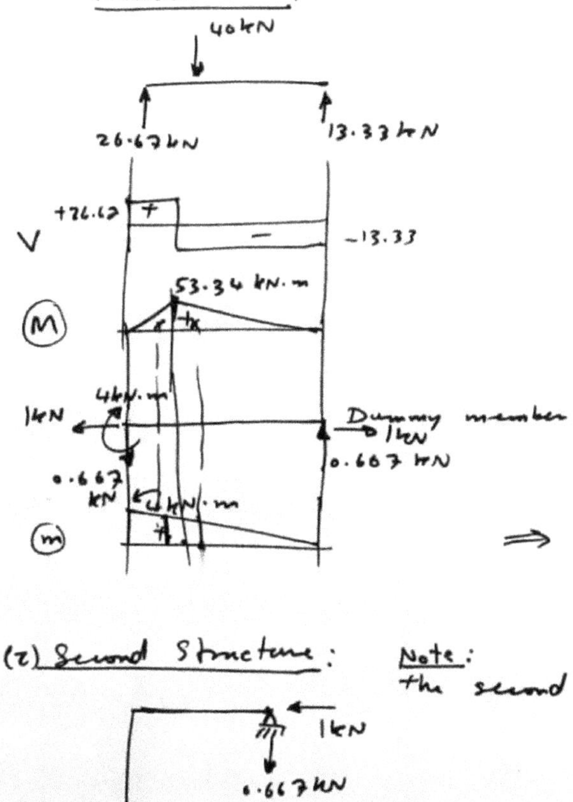

$$4 + \frac{1}{3}(2) = 4 + \frac{2}{3} = \frac{14}{3}$$

$$\int \frac{Mm\,dx}{EI} = \frac{1}{EI}\left[\frac{1}{2}(53.34)(2)\left(\frac{\frac{14}{3}}{6}\right)(4) + \frac{1}{2}(53.34)(4)\left(\frac{\frac{2}{3}(4)}{6}\right)(4)\right]$$

$$= \frac{355.6}{EI}$$

$$\Rightarrow \quad \delta_{A_0} = 0 + \frac{355.6}{EI} = \frac{355.6}{EI}$$

(to the left).

take $\delta_{A_0} = -\dfrac{355.6}{EI}$

(2) Second Structure:

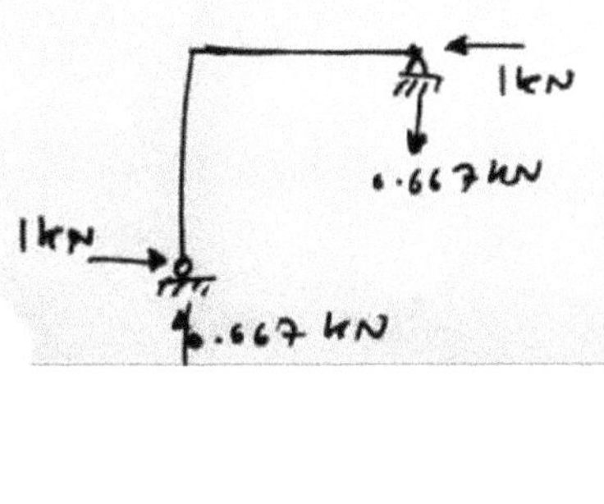

Note: the second structure is also the dummy structure

(H8)

Member AB :

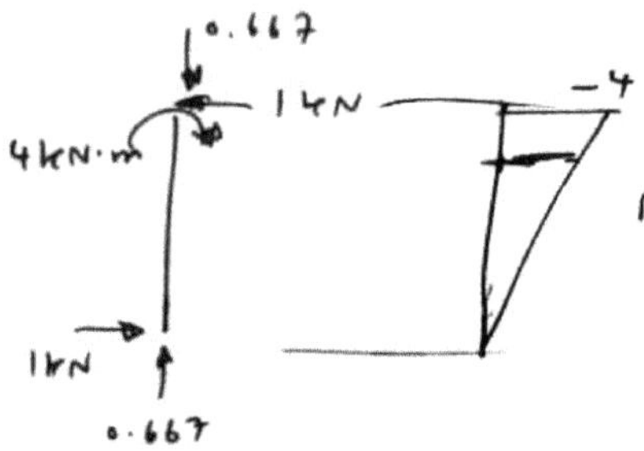

$$\int \frac{M m\,dx}{EI} = \frac{1}{EI}\left[\frac{1}{2}(4)(4)\left(\frac{2}{3}\right)\frac{(4)(4)}{4}\right]$$

$$= \frac{21.33}{EI}.$$

Member BC :

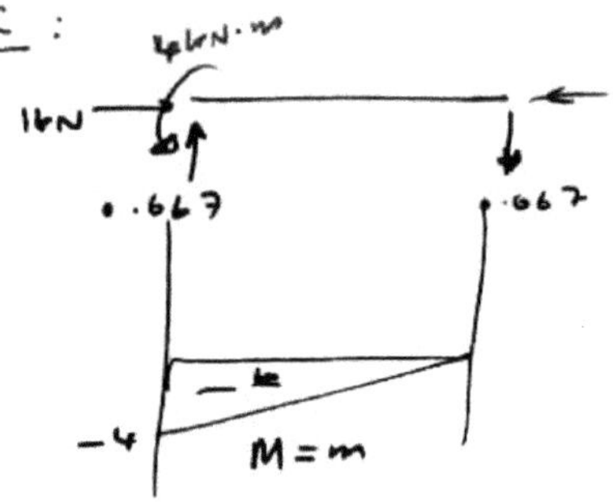

$$\int \frac{M m\,dx}{EI} = \frac{1}{EI}\left[\frac{1}{2}(6)(4)\left(\frac{2}{3}\right)\frac{(6)}{6}(4)\right]$$

$$= \frac{32}{EI}.$$

$$\therefore \quad \delta_{AA} = \frac{21.33}{EI} + \frac{32}{EI} = \frac{53.33}{EI} \quad (\text{to the right}).$$

(3) equation of consistent deformation :

$$\delta_{A_{(h)}} = \delta_{A0} + H_A\,\delta_{AA} = 0$$

$$-\frac{355.6}{EI} + H_A\left(\frac{53.33}{EI}\right) = 0$$

$$\Rightarrow \quad H_A = 6.67 \text{ kN} \rightarrow .$$

(4) equations of equilibrium :

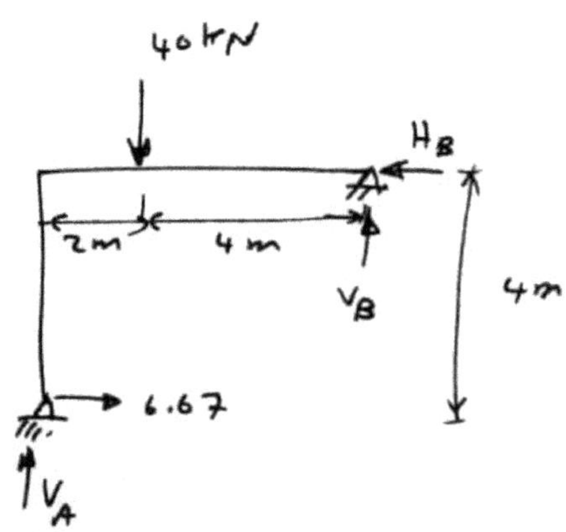

$\overset{+}{\rightarrow}\ \Sigma F_x = 0: \quad H_B = 6.67 \text{ kN} \leftarrow$

$+\circlearrowleft\ \Sigma M_A = 0:$

$$-40(2) + V_B(6) + 6.67(4) = 0$$

$$V_B = 8.89 \text{ kN} \uparrow .$$

$+\uparrow\ \Sigma F_y = 0:$

$$V_A + 8.89 - 40 = 0$$

$$V_A = 31.11 \text{ kN} \uparrow .$$

(H9)

B.M.D.

Member AB:

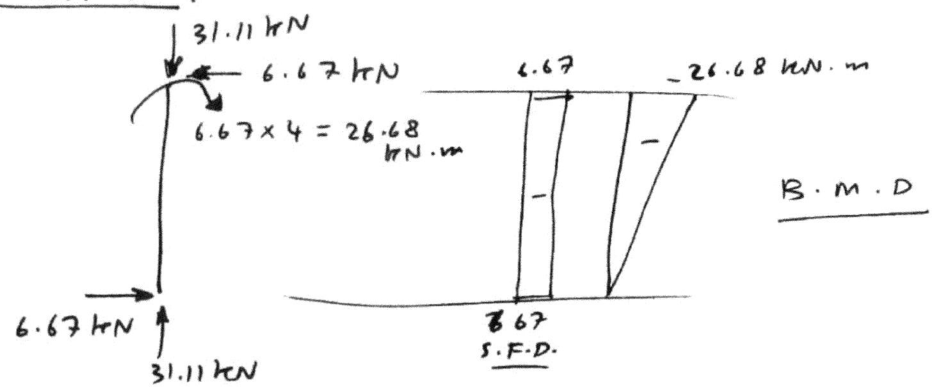

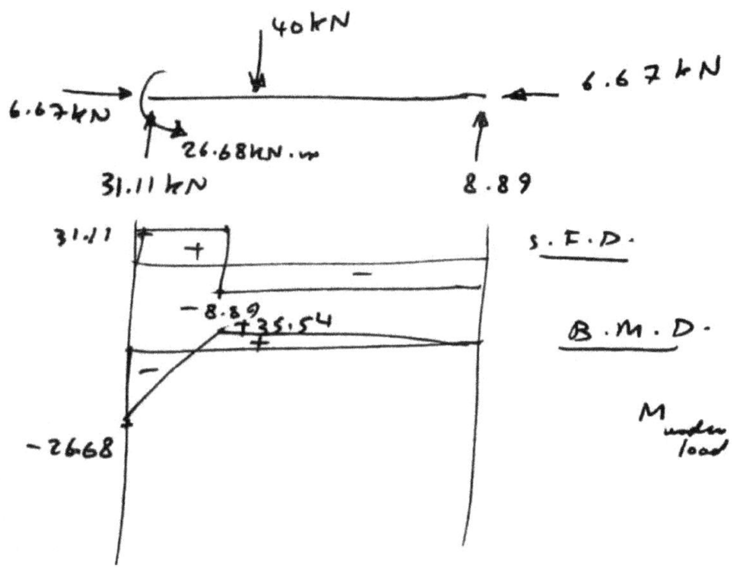

$$M_{\text{under load}} = -26.68 + 31.11(2)$$
$$= 35.54 \text{ kN.m}$$

Structural Analysis I – Exam-1- (Solutions)

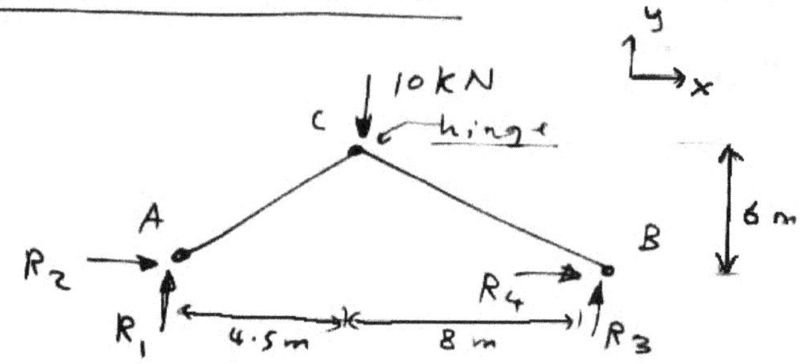

(1) Since the load is applied only to the joint, the structure is a truss.

$j = 3$, $2j = 6$.

$m = 2$, $r = 4$.

$\therefore$ $m + r = 6 = 2j$.

$\therefore$ the truss is stable and determinate.

We cannot find the reactions directly. So we use the method of joints. It is easy to start at joint C.

Joint C :

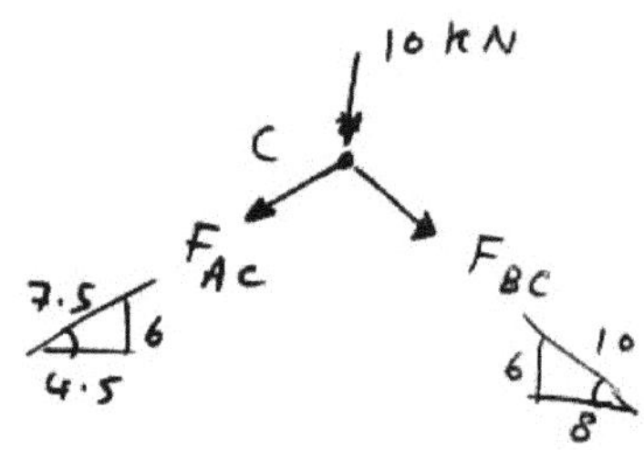

$\overset{+}{\rightarrow} \Sigma F_x = 0$:

$F_{BC}\left(\frac{8}{10}\right) - F_{AC}\left(\frac{4.5}{7.5}\right) = 0$ ——(1)

$+\uparrow \Sigma F_y = 0$:

$-10 - F_{BC}\left(\frac{6}{10}\right) - F_{AC}\left(\frac{6}{7.5}\right) = 0$ ——(2)

Solve the two equations simultaneously :

From eq. (1) : $F_{BC} = \left(\frac{10}{8}\right)\left(\frac{4.5}{7.5}\right) F_{AC} = 0.75\, F_{AC}$.

Substitute in eq. (2) :

$$-10 - \left(\frac{6}{10}\right)(0.75\, F_{AC}) - \frac{6}{7.5} F_{AC} = 0$$

$\therefore$ $F_{AC} = -8\text{ kN}$. $\Longrightarrow$ $F_{AC} = 8\text{ kN}$ (C) .

$F_{BC} = 0.75(-8) = -6\text{ kN}$ $\Longrightarrow$ $F_{BC} = 6\text{ kN}$ (C) .

Joint A :

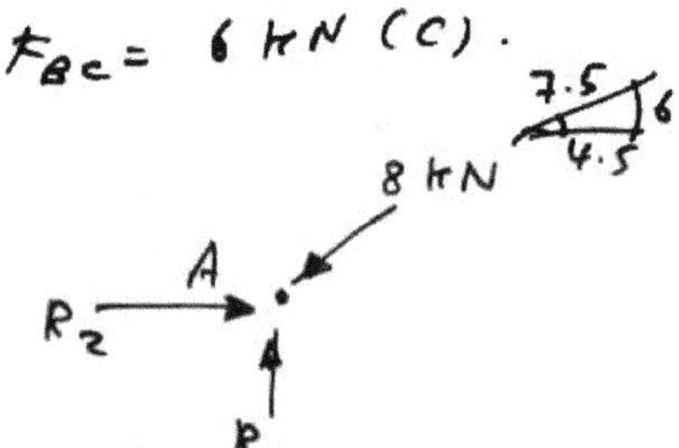

$\overset{+}{\rightarrow} \Sigma F_x = 0$: $R_2 - 8\left(\frac{4.5}{7.5}\right) = 0$

$\therefore$ $\boxed{R_2 = 4.8\text{ kN} \rightarrow}$

$+\uparrow \Sigma F_y = 0$: $R_1 - 8\left(\frac{6}{7.5}\right) = 0$

$\therefore$ $\boxed{R_1 = 6.4\text{ kN} \uparrow}$

continued →

(2)

Joint B:

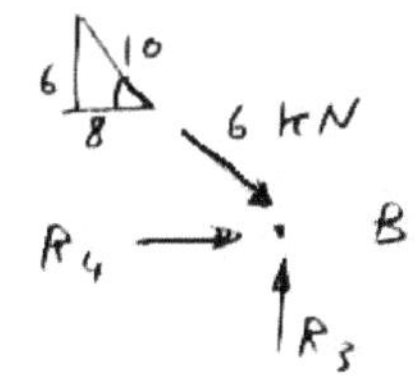

$\overset{+}{\rightarrow} \Sigma F_x = 0: \quad R_4 + 6\left(\frac{8}{10}\right) = 0$

$R_4 = -4.8$ kN.

$\therefore$ $\boxed{R_4 = 4.8 \text{ kN} \leftarrow}$

$+\uparrow \Sigma F_y = 0: \quad R_3 - 6\left(\frac{6}{10}\right) = 0$

$\therefore$ $\boxed{R_3 = 3.6 \text{ kN} \uparrow}$

---

(2)

(a) Number of Unknowns (Reactions) = 3

Number of Equations = 3 + 1 = 4

$\therefore$ No. of Unknowns < No. of Equations

$\therefore$ The beam is $\boxed{\text{unstable.}}$

(b) No. of Unknowns (Reactions) = 5

No. of Equations = 3 + 2 = 5

$\therefore$ No. of Unknowns = No. of Equations.

$\therefore$ the beam is $\boxed{\text{stable}}$ and $\boxed{\text{determinate}}$.

(c) $j = 5$. $\therefore$ No. of Equations $= 2j = 10$.

$m = 7$

$r = 3$

No. of Unknowns $= m + r = 7 + 3 = 10$.

$\therefore$ $m + r = 2j$.

But the truss has three parallel reactions.

$\therefore$ The truss is $\boxed{\text{unstable}}$.

(d) No. of Reactions $r = 4 > 3$

$\therefore$ the truss is externally indeterminate.

However, $j = 18$

No. of Equations $= 2j = 36$.

$r = 4$

$m = 32$

No. of Unknowns $= m + r = 36 \Rightarrow m + r = 2j$.

$\therefore$ The truss is $\boxed{\text{stable}}$ and $\boxed{\text{determinate.}}$

---

③ Calculate the reactions:

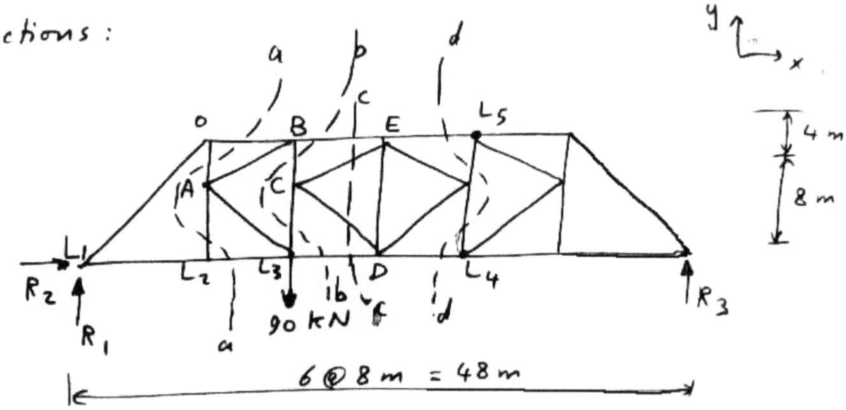

$\xrightarrow{+} \Sigma F_x = 0: \quad R_2 = 0.$

$+\circlearrowleft \Sigma M_{L_1} = 0:$

$R_3(48) - 90(16) = 0$

$R_3 = 30 \text{ kN} \uparrow.$

$+\uparrow \Sigma F_y = 0:$

$R_1 + 30 - 90 = 0$

$R_1 = 60 \text{ kN} \uparrow.$

Section a-a:

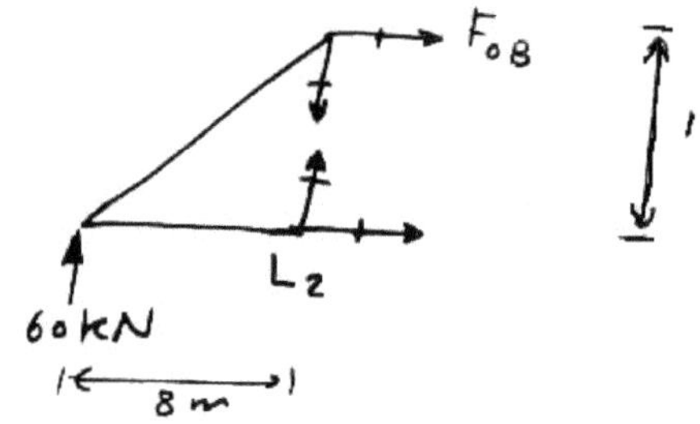

$+\circlearrowleft \Sigma M_{L_2} = 0:$

$-F_{OB}(12) - 60(8) = 0$

$F_{OB} = -40 \text{ kN}.$

$\therefore \boxed{F_{OB} = 40 \text{ kN}, (C).}$

Section b-b:

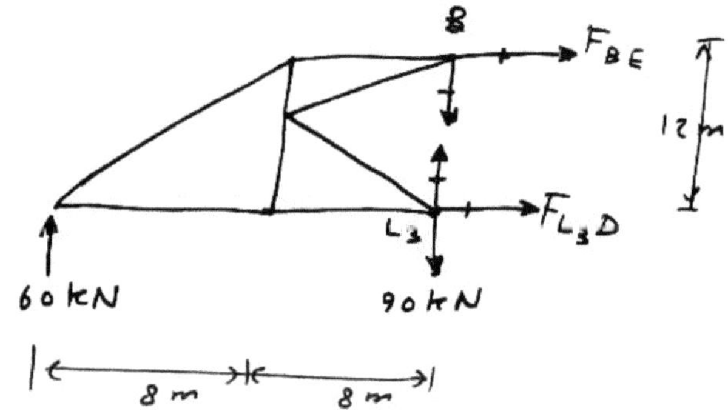

$+\circlearrowleft \Sigma M_{L_3} = 0:$

$-F_{BE}(12) - 60(16) = 0$

$\therefore F_{BE} = -80 \text{ kN}.$

$\therefore \boxed{F_{BE} = 80 \text{ kN}, (C).}$

Joint B:

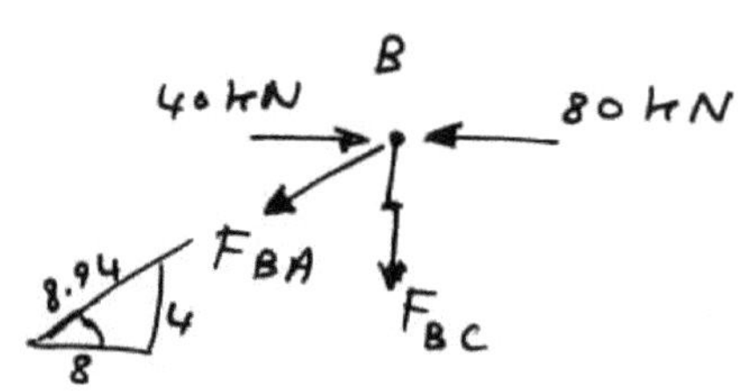

$\xrightarrow{+} \Sigma F_x = 0:$

$-80 + 40 - F_{BA}\left(\frac{8}{8.94}\right) = 0$

$F_{BA} = -44.7 \text{ kN}.$

$+\uparrow \Sigma F_y = 0:$

$-F_{BC} - (-44.7)\left(\frac{4}{8.94}\right) = 0$

$\therefore \boxed{F_{BC} = +20 \text{ kN}, (T).}$

continued →

(4)

Section b-b (again):

$+\circlearrowleft \sum M_B = 0:$

$$F_{L_3D}(12) - 60(16) = 0$$

$$\therefore \quad F_{L_3D} = +80 \text{ kN}.$$

Section c-c:

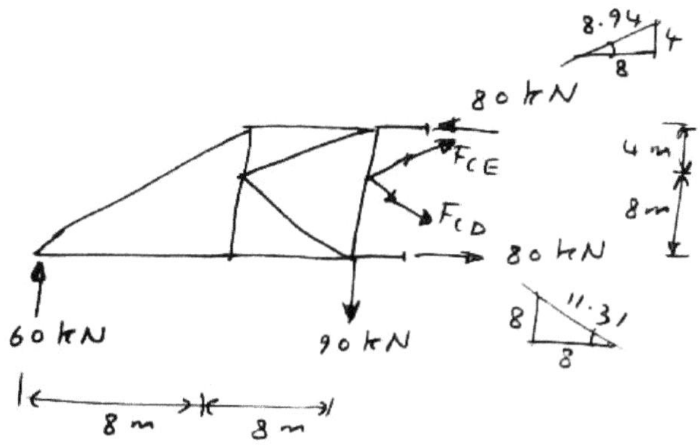

$\overset{+}{\rightarrow} \sum F_x = 0:$

$$\cancel{80} - \cancel{80} + F_{CE}\left(\frac{8}{8.94}\right) + F_{CD}\left(\frac{8}{11.31}\right) = 0$$

$$\therefore \; F_{CD} = \frac{-11.31}{8.94} F_{CE}$$

$$= -1.265 \, F_{CE} \quad \text{——— (1)}$$

$+\uparrow \sum F_y = 0:$

$$60 - 90 + F_{CE}\left(\frac{4}{8.94}\right) - F_{CD}\left(\frac{8}{11.31}\right) = 0$$

substitute for F_{CD} from eq. (1):

$$-30 + F_{CE}\left(\frac{4}{8.94}\right) - \left(\frac{8}{11.31}\right)(-1.265 \, F_{CE}) = 0$$

$$\boxed{F_{CE} = 22.35 \text{ kN, (T).}}$$

Section d-d:

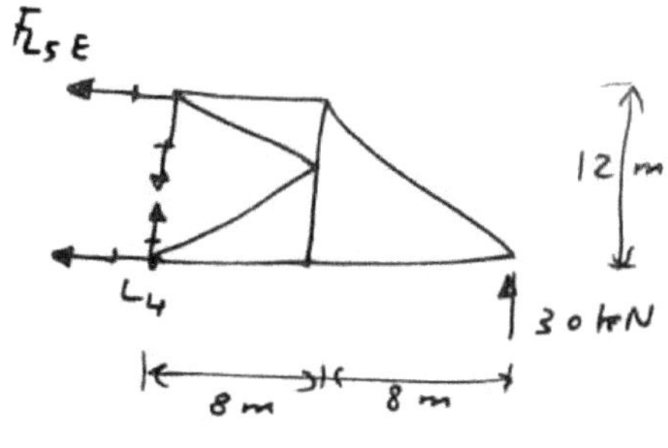

$+\circlearrowleft \sum M_{L_4} = 0:$

$$F_{L_5E}(12) + 30(16) = 0$$

$$F_{L_5E} = -40 \text{ kN}.$$

Joint E:

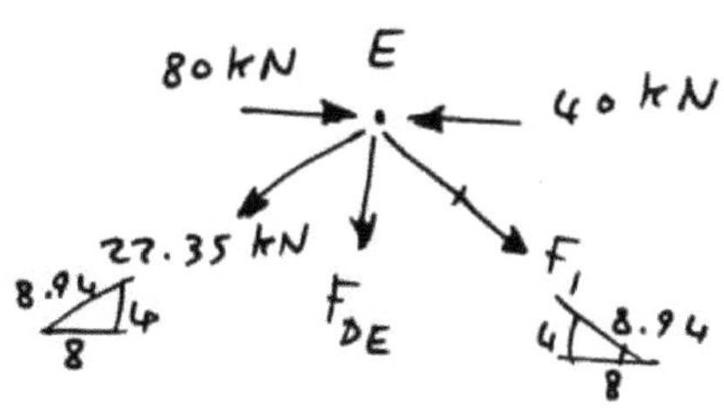

$\overset{+}{\rightarrow} \sum F_x = 0:$

$$80 - 40 + F_1\left(\frac{8}{8.94}\right) - 22.35\left(\frac{8}{8.94}\right) = 0$$

$$\therefore \; F_1 = -22.35 \text{ kN}.$$

$+\uparrow \sum F_y = 0:$

$$-\cancel{(-22.35)}\left(\frac{4}{8.94}\right) - F_{DE} - \cancel{22.35}\left(\frac{4}{8.94}\right) = 0$$

$$\boxed{F_{DE} = 0} \Rightarrow \text{zero-force member.}$$

(5)

(4) Calculate the reactions:

Separate the frame at the hinge:

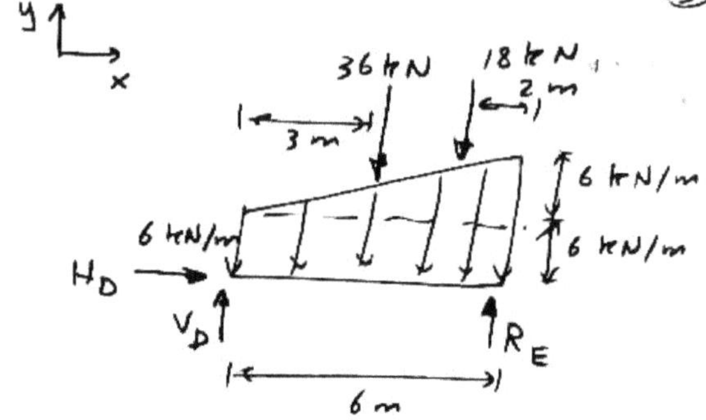

Member DE:

$+\circlearrowleft\ \Sigma M_D = 0:$

$$R_E(6) - 36(3) - 18(4) = 0$$

$\therefore$ $\boxed{R_E = 30\ kN\ \uparrow}$

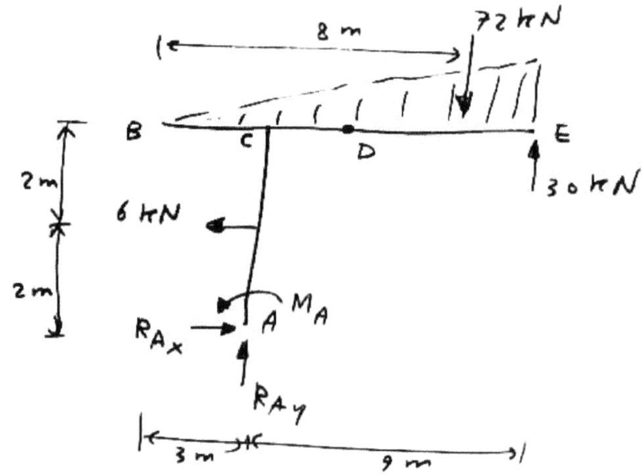

Entire Frame:

$\xrightarrow{+}\ \Sigma F_x = 0:$

$$R_{Ax} - 6 = 0$$

$\boxed{R_{Ax} = 6\ kN \rightarrow}$

$+\uparrow\ \Sigma F_y = 0:$

$$R_{Ay} + 30 - 72 = 0$$

$\boxed{R_{Ay} = 42\ kN\ \uparrow}$

$+\circlearrowleft\ \Sigma M_A = 0:$

$$M_A + 6(2) + 30(9) - 72(5) = 0$$

$\therefore$ $\boxed{M_A = +78\ kN\cdot m}$

Member AC:

$+\circlearrowleft\ \Sigma M_c = 0:$

$$M_c + 78$$
$$+ 6(4) - 6(2) = 0$$

$\therefore$ $M_c = -90\ kN\cdot m$

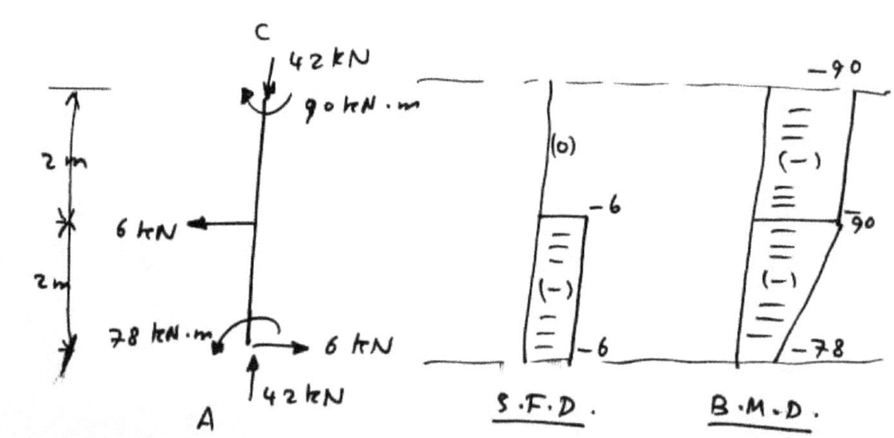

continued →

(6)

Member BCDE :

$V_C^- = -\frac{1}{2}(3)(3) = -4.5$ kN.

$V_C^+ = -4.5 + 42 = 37.5$ kN.

$V_E = 37.5 - \frac{1}{2}(3+12)(9) = -30$ kN.

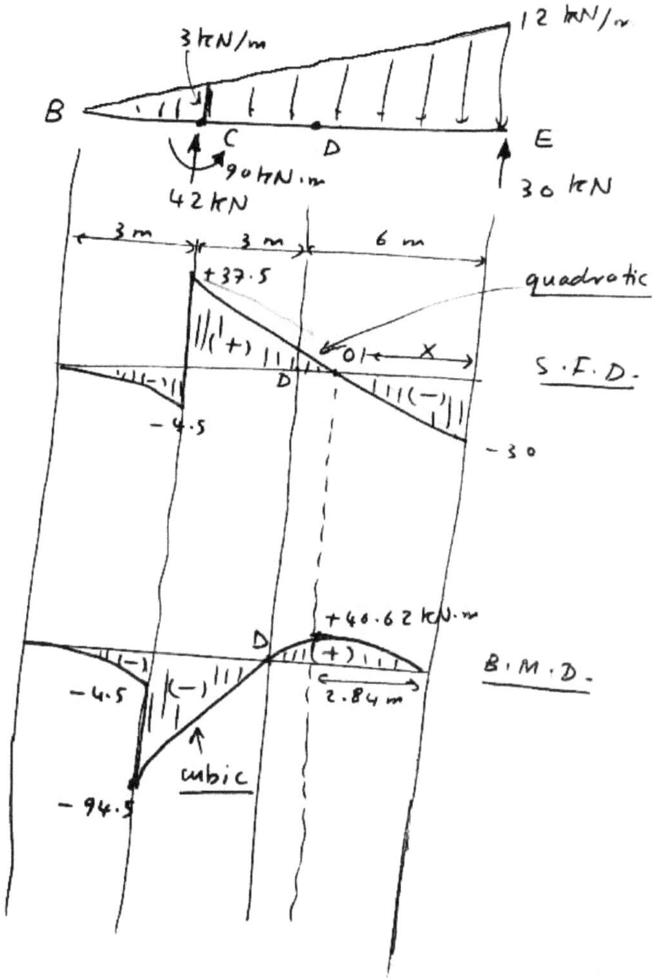

Locate point O of zero shear:

Take a section at O, distance x from E.

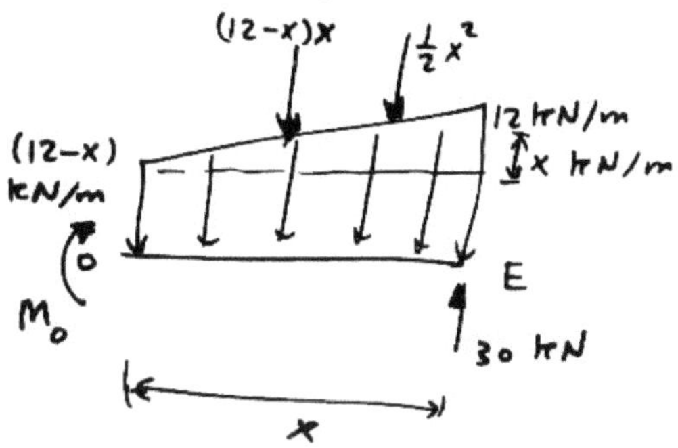

V = 0 at point O.

$\Sigma F_y = 0$:

$30 - (12-x)x - \frac{1}{2}x^2 = 0$

$30 - 12x + x^2 - \frac{1}{2}x^2 = 0$

$30 - 12x + \frac{1}{2}x^2 = 0$

$x^2 - 24x + 60 = 0$

$x = \frac{24 \pm \sqrt{(24)^2 - 4(1)(60)}}{2(1)} = \frac{24 \pm 18.33}{2}$ → 21.16 > 12 m ✗ ; → $\boxed{2.84\text{ m}}$ ✓

$\therefore \boxed{x = 2.84\text{ m}}$

$+\circlearrowleft \Sigma M_O = 0$:

$-M_O + 30(2.84) - (12-2.84)(2.84)\left(\frac{2.84}{2}\right) - \frac{1}{2}(2.84)^2 \cdot \frac{2}{3}(2.84) = 0$

$\therefore M_O = +40.62$ kN.m.

Joint C : $M_C^- = -\frac{1}{2}(3)(3)\cdot\frac{1}{3}(3) = -4.5$ kN.m

$M_C^+ = -4.5 - 90 = -94.5$ kN.m

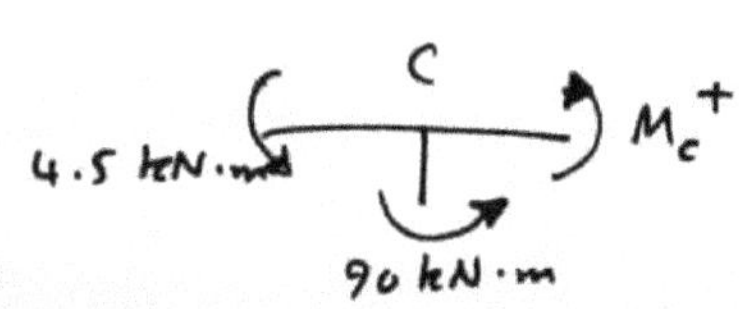

Structural Analysis (1)

Exam -1- / Solutions

(1) Calculate the forces in members DG, EG, EI, AB and MN of the K-truss shown in the figure. Indicate whether each member is in tension or compression.

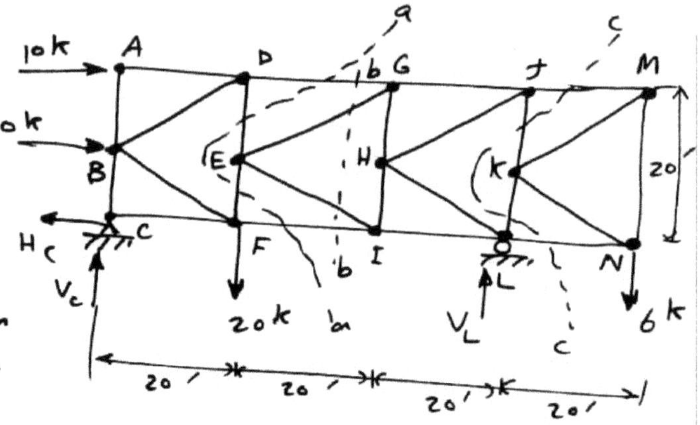

Solution:

(1) Calculate the reaction:

$$+\circlearrowleft \Sigma M_C = 0: \quad V_L(60) - 20(20) - 6(80) - 10(10) - 10(20) = 0$$

$$\Rightarrow V_L = 19.67\ k \uparrow .$$

$$+\uparrow \Sigma F_y = 0: \quad V_C - 20 + (19.67) - 6 = 0 \Rightarrow V_C = 6.33\ k \uparrow .$$

$$\xrightarrow{+} \Sigma F_x = 0: \quad -H_C + 10 + 10 = 0 \Rightarrow H_C = 20k \leftarrow .$$

Section a-a:

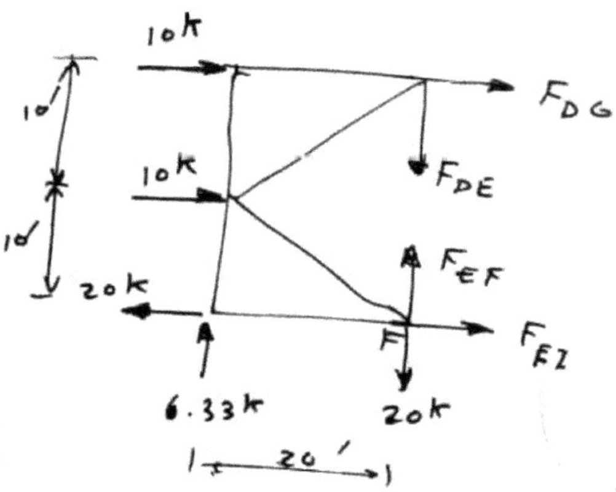

$$+\circlearrowleft \Sigma M_F = 0:$$

$$-F_{DG}(20) - 10(20) - 10(10) - 6.33(20) = 0$$

$$\boxed{F_{DG} = -21.33k\ (C).}$$

Joint E:

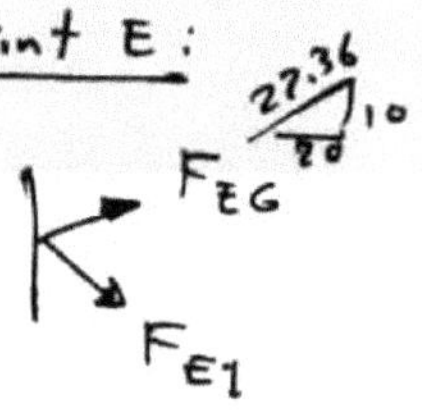

$$\xrightarrow{+} \Sigma F_x = 0: \quad F_{EG}\left(\frac{20}{27.36}\right) + F_{EI}\left(\frac{20}{27.36}\right) = 0 \Rightarrow F_{EG} = -F_{EI} .$$

Section b-b:

$$+\uparrow \Sigma F_y = 0:$$

$$F_{EG}\left(\frac{10}{27.36}\right) - F_{EI}\left(\frac{10}{27.36}\right) + 6.33 - 20 = 0$$

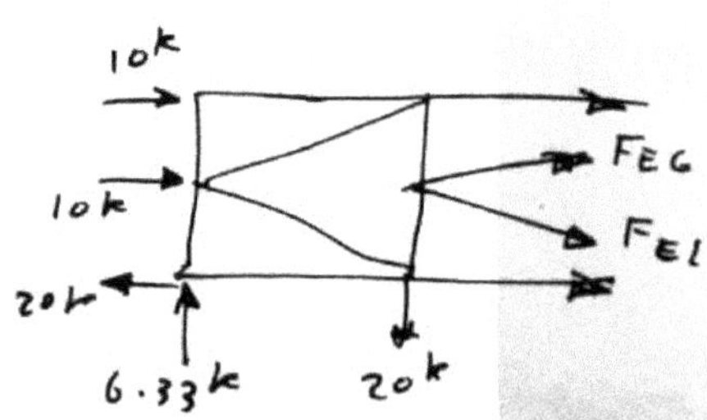

But $F_{EG} = -F_{EI}$

$$\Rightarrow \quad F_{EG}\left(\frac{10}{22.36}\right) + F_{EG}\left(\frac{10}{22.36}\right) = 20 - 6.33$$

$$2F_{EG}\left(\frac{10}{22.36}\right) = 13.67 \quad \Rightarrow \boxed{F_{EG} = 15.28\text{ k (T)}.}$$

$$\Rightarrow \boxed{F_{EI} = -15.28\text{ k (C)}.}$$

Joint A:

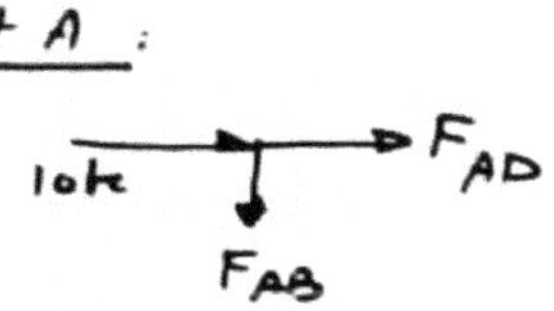

$+\uparrow \sum F_y = 0: \quad \Rightarrow \boxed{F_{AB} = 0}$

Section c–c:

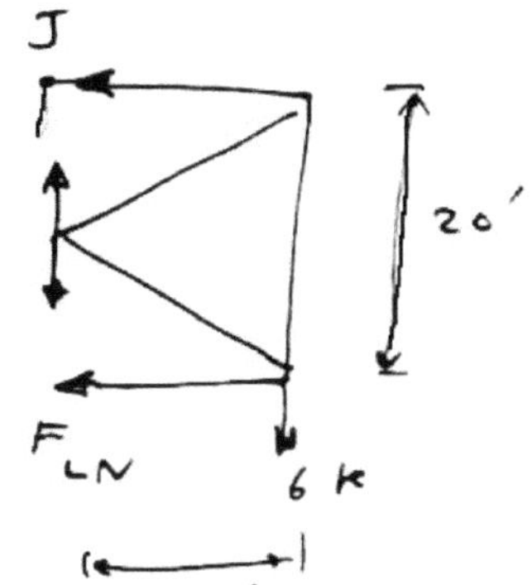

$+\circlearrowleft \sum M_J = 0:$

$$-F_{LN}(20) - 6(20) = 0$$

$$F_{LN} = -6\text{ k (C)}.$$

Joint N:

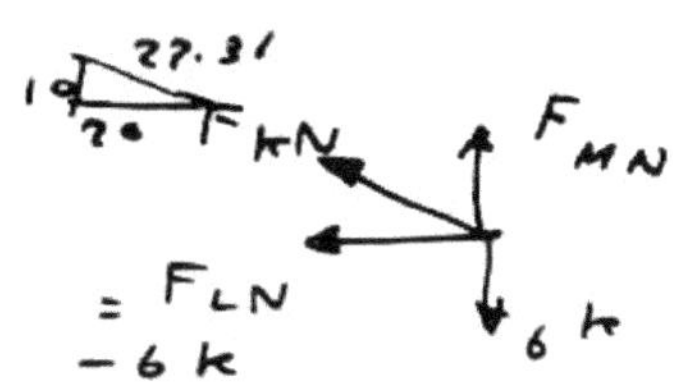

$\xrightarrow{+} \sum F_x = 0:$

$$6 - F_{KN}\left(\frac{20}{22.36}\right) = 0$$

$$\Rightarrow F_{KN} = 6.71\text{ kN (T)}.$$

$+\uparrow \sum F_y = 0: \quad F_{MN} - 6 + (6.71)\left(\frac{10}{22.36}\right) = 0$

$$\Rightarrow \boxed{F_{MN} = +3\text{ k (T)}.}$$

---

(2) Draw the shear force and bending moment diagrams for the compound beam shown in the figure.

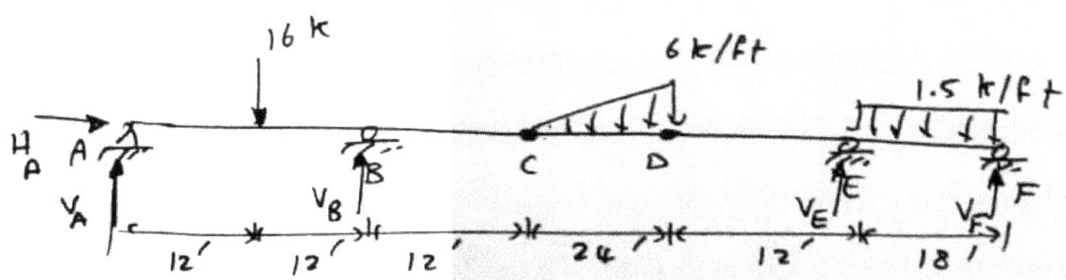

Solution:

Member CD:

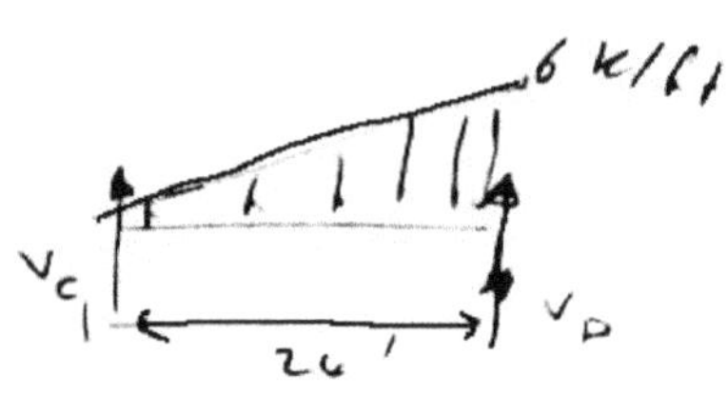

$+\circlearrowleft \sum M_C = 0:$

$$V_D(24) - \frac{1}{2}(6)(24)\cdot\frac{2}{3}(24) = 0$$

$$V_D = 48\ k.$$

$+\uparrow \sum F_y = 0: \quad V_C + (48) - \frac{1}{2}(6)(24) = 0 \Rightarrow V_C = 24\ k.$

Member ABC:

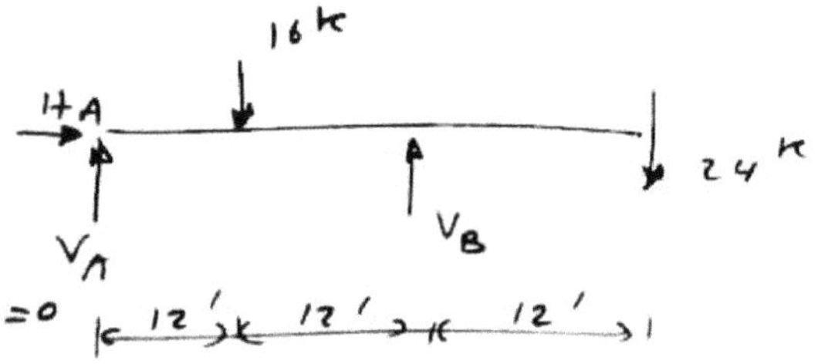

$H_A = 0.$ ✓

$+\circlearrowleft \sum M_A = 0:$

$$-24(36) - 16(12) + V_B(24) = 0$$

$$V_B = 44\ k.$$

$+\uparrow \sum F_y = 0: \quad V_A + (44) - 16 - 24 = 0 \Rightarrow V_A = -4\ k.$

$\therefore V_A = 4\ k \downarrow.$

Member EF:

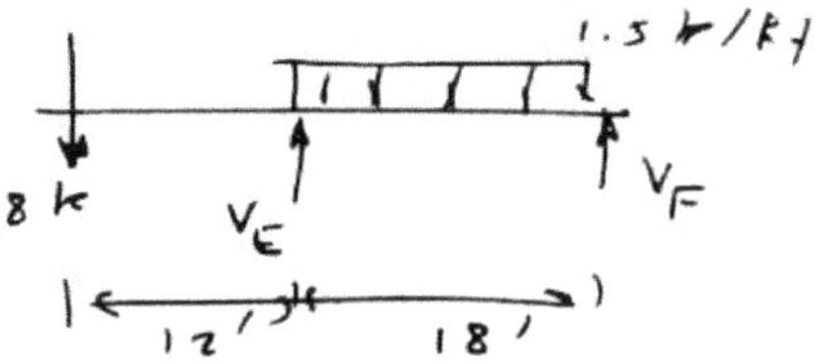

$+\circlearrowleft \sum M_F = 0:$

$$1.5(18)(9) + 48(30) - V_E(18) = 0$$

$$V_E = 93.5\ k.$$

$+\uparrow \sum F_y = 0: \quad -48 + (93.5) + V_F - 1.5(18) = 0$

$\Rightarrow V_F = -18.5\ k \Rightarrow V_F = 18.5\ k \downarrow.$

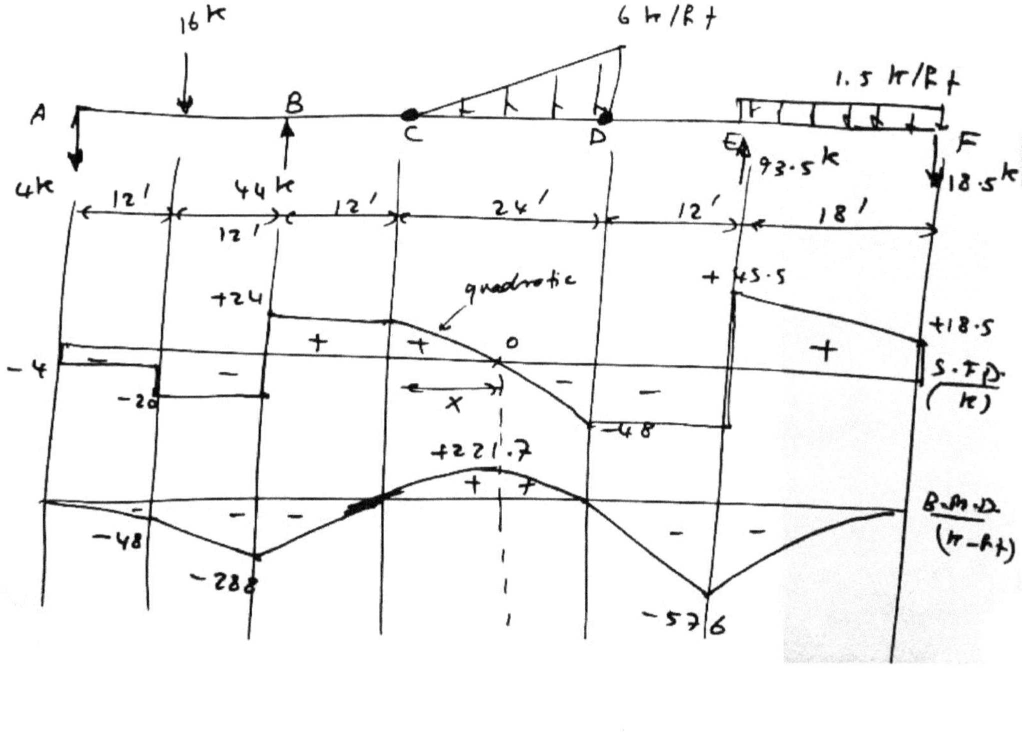

L4

Locate point O of zero shear:

$$\frac{x}{y} = \frac{24}{6} \Rightarrow y = \frac{1}{4}x.$$

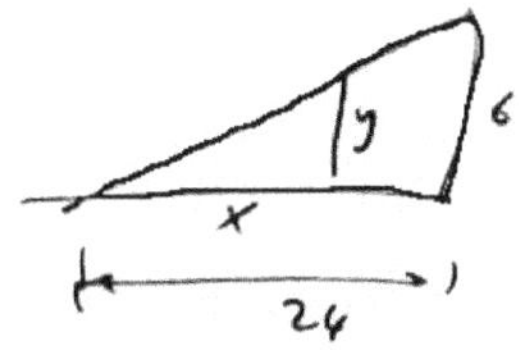

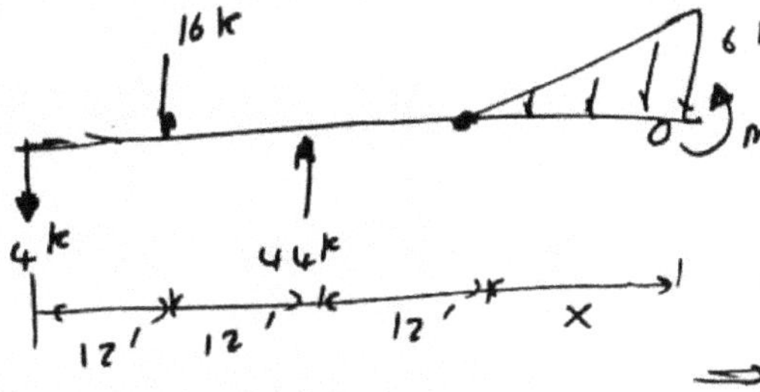

$+\uparrow \Sigma F_y = 0:$

$$-4 - 16 + 44 - \tfrac{1}{2}(x)(\tfrac{1}{4}x) = 0$$

$$\Rightarrow \frac{x^2}{8} = 24 \Rightarrow x = 13.86'.$$

$$M_o = -4(36+13.86) - 16(24+13.86) + 44(12+13.86) - \tfrac{1}{2}(13.86)\cdot\tfrac{1}{4}(13.86)\cdot\tfrac{1}{3}(13.86) = +\underline{\underline{221.7}}\ \text{k-ft}.$$

---

③ Determine the slope (rotation) at A and B, and the deflection at A in the cantilever beam shown in the figure. Take $E = 1.4\times10^6$ psi, $I = 1200$ in^4.

Solution:

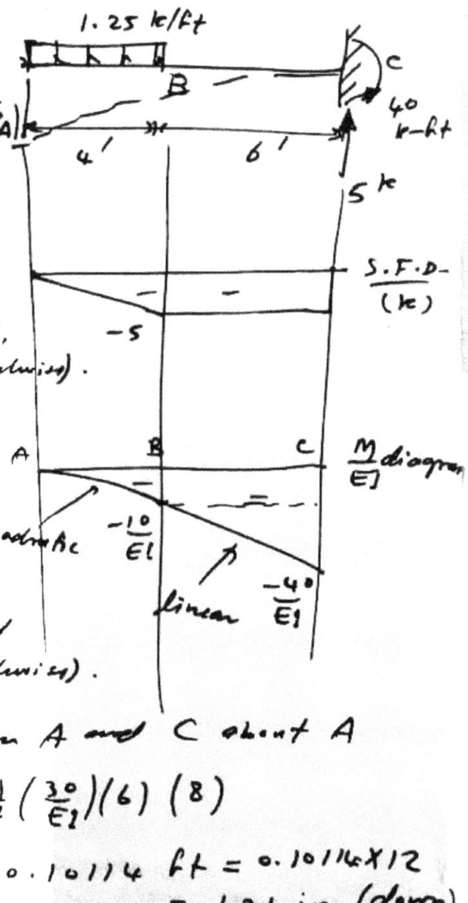

θ_A = area of $\frac{M}{EI}$ diagram between A and C

$$= \tfrac{1}{3}(4)\left(\tfrac{10}{EI}\right) + \tfrac{1}{2}\left(\tfrac{10}{EI} + \tfrac{40}{EI}\right)(6)$$

$$= \frac{163.33}{EI}$$

$$= \frac{163.33 \times (12)^2}{1.4\times10^6 \times 10^{-3} \times 1200} = 0.014 \text{ rad, (counterclockwise).}$$

θ_B = area of $\frac{M}{EI}$ diagram between B and C

$$= \tfrac{1}{2}\left(\tfrac{10}{EI} + \tfrac{40}{EI}\right)(6) = \frac{150}{EI}$$

$$= \frac{150 \times (12)^2}{1.4\times10^6 \times 10^{-3} \times 1200} = 0.013 \text{ rad (counterclockwise).}$$

δ_A = moment of $\frac{M}{EI}$ diagram between A and C about A

$$= \tfrac{1}{3}\left(\tfrac{10}{EI}\right)(4)\left(\tfrac{3}{4}\right)(4) + \left(\tfrac{10}{EI}\right)(6)(7) + \tfrac{1}{2}\left(\tfrac{30}{EI}\right)(6)(8)$$

$$= \frac{1180}{EI} = \frac{1180 \times (12)^2}{1.4\times10^6 \times 10^{-3} \times 1200} = 0.10114 \text{ ft} = 0.10114 \times 12$$

$$= \underline{1.21} \text{ in. (down)}$$

Structural Analysis (1)

Exam -3- / Solutions

① Draw the influence line for the force in member AB of the truss shown in the figure.

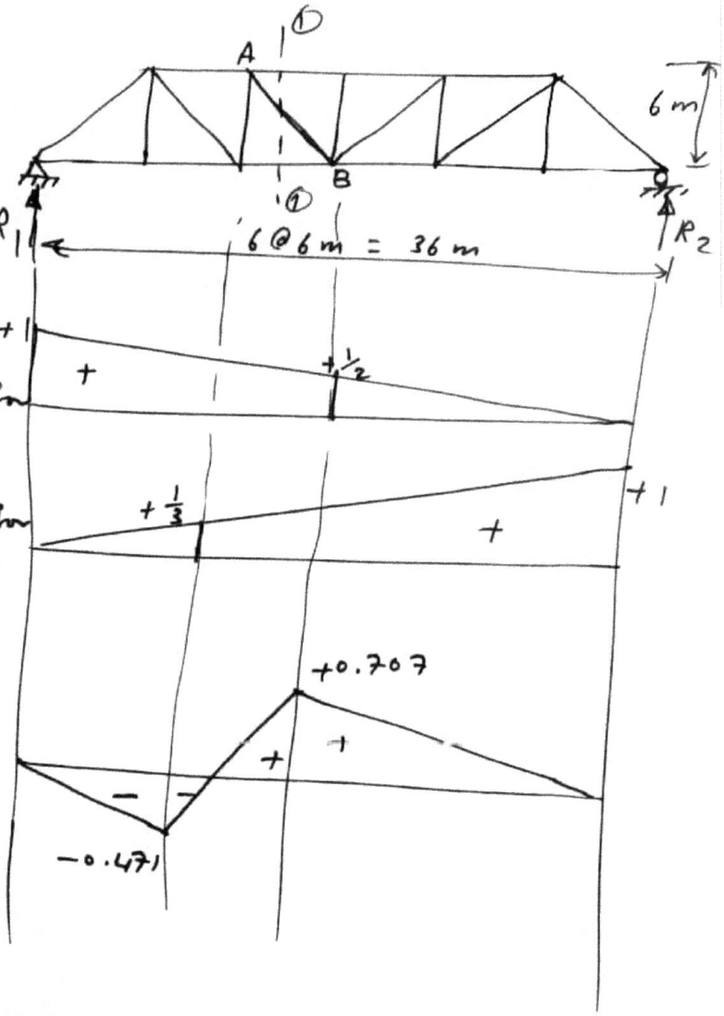

Solution:

Case (1): Load to the left of A:

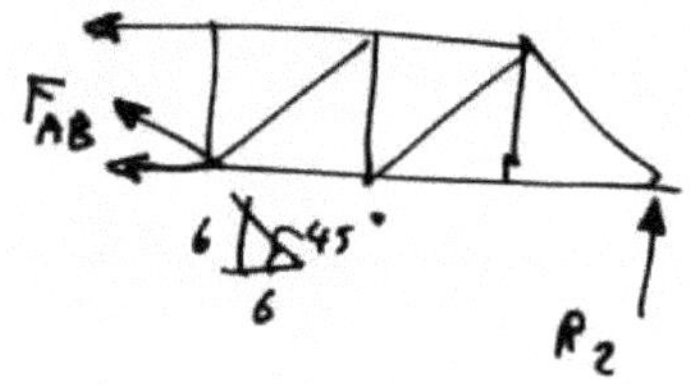

$+\uparrow \Sigma F_y = 0:$

$$F_{AB} \sin 45^\circ + R_2 = 0$$

$$\boxed{F_{AB} = -1.414\, R_2}$$

Case (2): Load to the right of A:

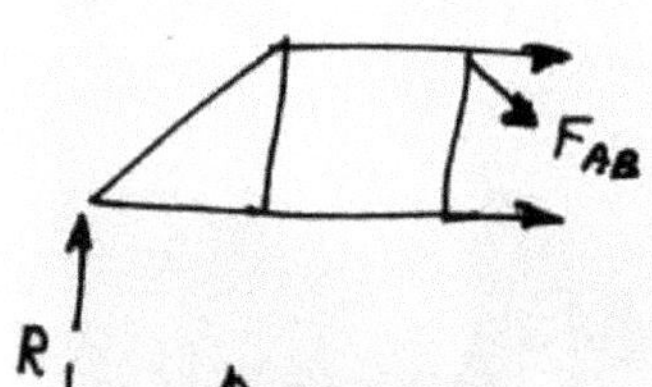

$+\uparrow \Sigma F_y = 0:$

$$R_1 - F_{AB} \sin 45^\circ = 0$$

$$\boxed{F_{AB} = +1.414\, R_1}$$

(2) The simply-supported beam shown in the figure is subjected to a set of two concentrated moving loads. Determine the position of the loads and the resulting value of the absolute maximum moment in the beam.

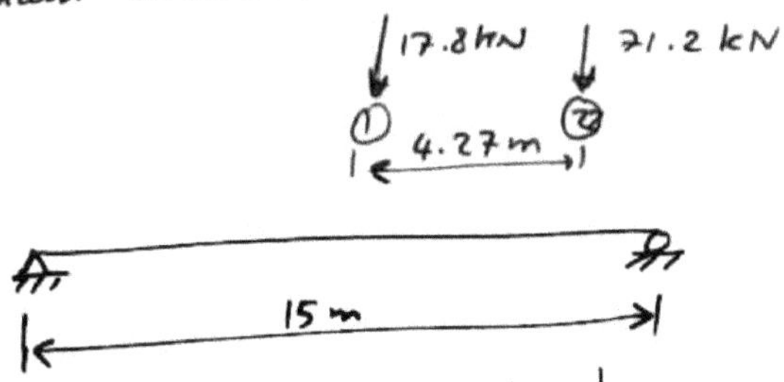

Solution :

Absolute maximum moment occurs under load (2).

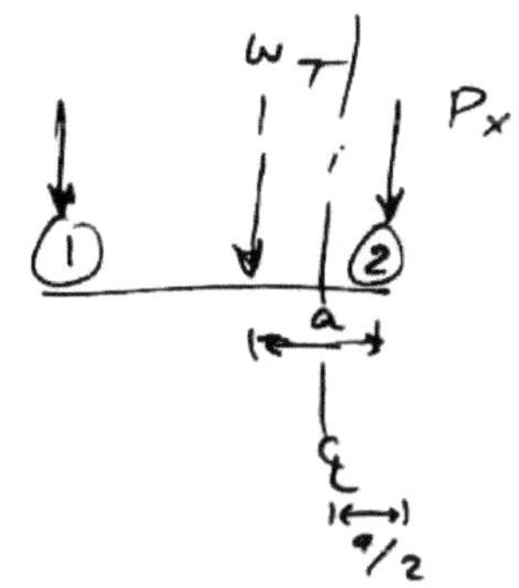

$$\bar{x} = \frac{17.8(0) + 71.2(4.27)}{(17.8 + 71.2)} = 3.416 \text{ m.}$$

$$a = 4.27 - 3.416 = 0.854 \text{ m.}$$

$$x = \frac{L}{2} + \frac{a}{2} = \frac{15}{2} + \frac{0.854}{2} = 7.927 \text{ m.}$$

$$R_1 = 71.2\left(\frac{7.073}{15}\right) + 17.8\left(\frac{11.343}{15}\right)$$

$$= 47.03 \text{ kN.}$$

$$R_2 = 17.8 + 71.2 - 47.03$$

$$= 41.97 \text{ kN.}$$

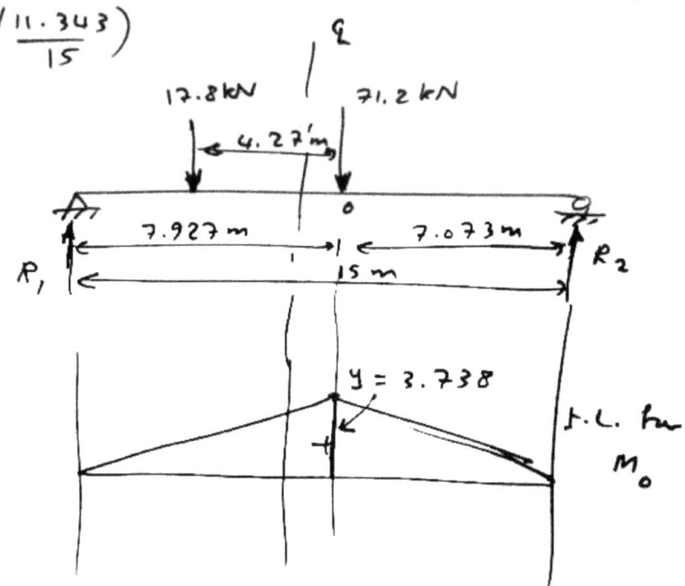

$$M_{o\,(max.)} = 47.03(7.927) - 17.8(4.27) = 296.8 \text{ kN.m.}$$

Absolute Max. Moment

$$= \boxed{296.8 \text{ kN.m.}}$$

Alternative Method:

Using the influence line:

$$M_{o\,(max.)} = 71.2(3.738) + 17.8\left(\frac{3.657}{7.927}\right)(3.738)$$

$$= \boxed{296.8 \text{ kN.m.}}$$

13

(3) The shapes of the two segments of the three-hinged arch are described by the functions shown. The two supports are not on the same level. Determine:

(a) the horizontal and vertical components of reaction at A and C.

(b) the moment in the arch at point E.

Solution:

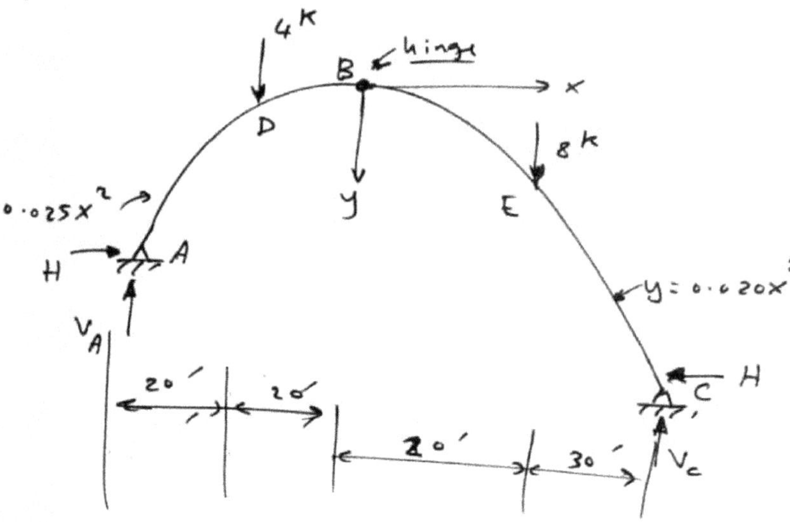

(a)

$y_A = 0.025(40)^2 = 40'$.

$y_c = 0.020(50)^2 = 50'$.

$+\circlearrowright \Sigma M_A = 0: \quad V_c(90) - H(10) - 8(60) - 4(20) = 0$

$$10H - 90V_c = -560 \quad (1)$$

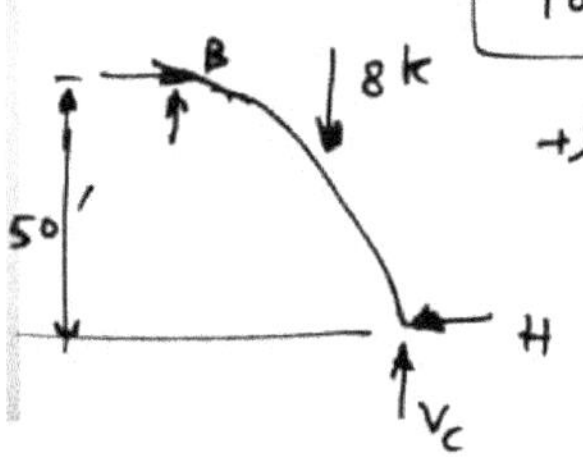

$+\circlearrowright \Sigma M_B = 0:$

$V_c(50) - H(50) - 8(20) = 0$

$$50H - 50V_c = -160 \quad (2)$$

Solve eqs. (1) & (2) simultaneously for V and H?

$-50H + 450V_c = 2800$

$50H - 50V_c = -160$

$\Rightarrow V_c = +6.6\ k\uparrow$.

$\Rightarrow H = 3.4\ k$.

For the entire arch:

$+\uparrow \Sigma F_y = 0: \quad V_A = 4 + 8 - 6.6 = 5.4\ k\uparrow$.

$y_E = 0.020(20)^2 = 8'$.

$y_1 = 50 - 8 = 42'$.

$\Rightarrow M_E = 6.6(30) - 3.4(42) = +55.2\ k\text{-}ft$.

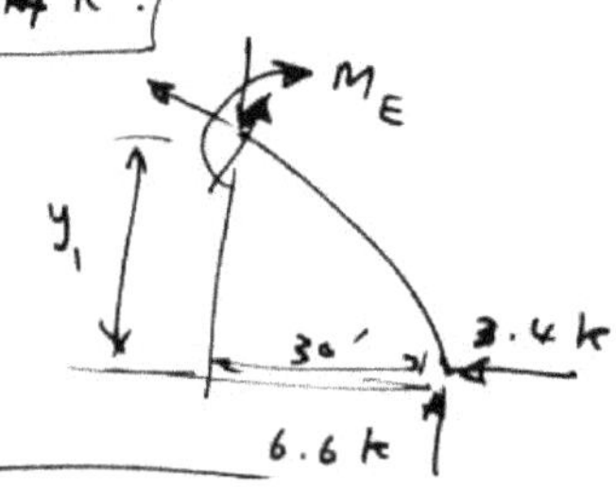

Structural Analysis - I - / Exam - 3 - Solutions

(1) Consider the compound beam shown in the figure. Draw the influence lines for:

(a) the reactions at support A.

(b) the shear force and moment at point B.

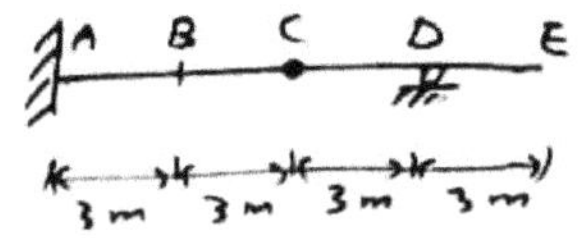

Solution:

(a) Reactions at A:

Case (1): Unit load to the left of the hinge at C:

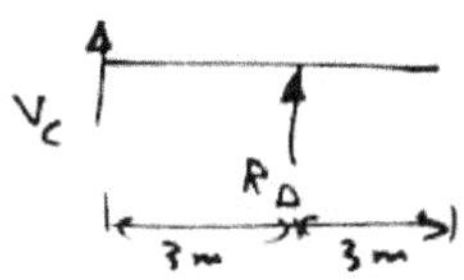

$+\circlearrowleft \Sigma M_c = 0: \; 3R_D = 0 \Rightarrow R_D = 0.$

For the entire beam:

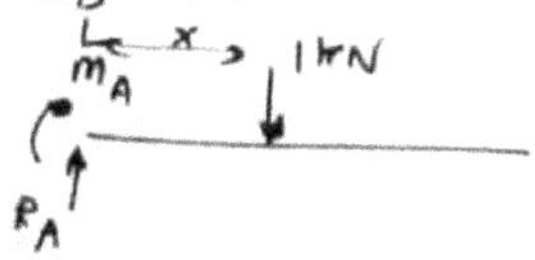

$+\uparrow \Sigma F_y = 0: \quad R_A = 1$

$+\circlearrowleft \Sigma M_A = 0: \; M_A = -x,$

$0 \le x \le 6$ from the left end.

Case (2): Unit load to the right of the hinge at C:

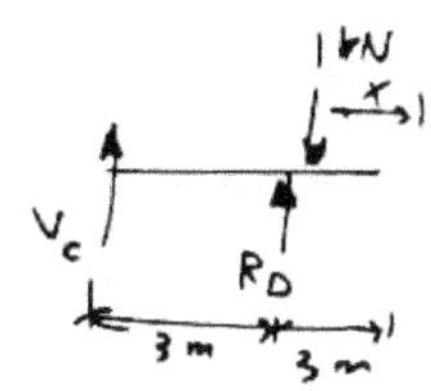

$+\circlearrowleft \Sigma M_c = 0: \quad 3R_D - 1(6-x) = 0$

$R_D = \frac{6-x}{3} = 2 - \frac{1}{3}x, \quad 0 \le x \le 6$

For the entire beam:

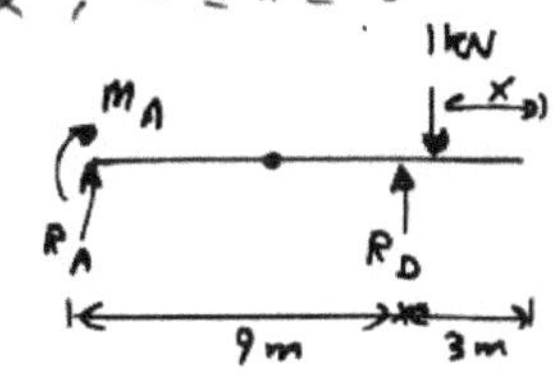

$+\uparrow \Sigma F_y = 0: \quad R_A = 1 - R_D$

$= 1 - (2 - \frac{1}{3}x)$

$= \frac{1}{3}x - 1$

$+\circlearrowleft \Sigma M_A = 0: M_A = 9R_D - 1(12 - x)$

$\therefore \; M_A = 9\left(\frac{6-x}{3}\right) - (12 - x)$

$= 18 - 3x - 12 + x$

$= 6 - 2x, \quad 0 \le x \le 6$ from the right end.

(b) Shear and Moment at B:

Case (1): Unit load to the left of B:

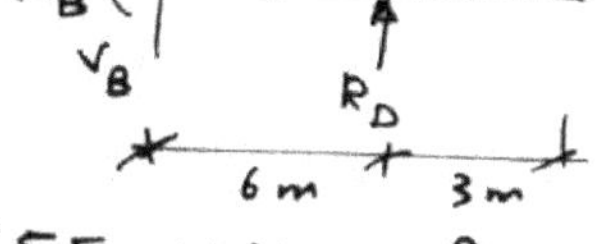

$+\uparrow \Sigma F_y = 0: \; V_B = -R_D$

$+\circlearrowleft \Sigma M_B = 0: \quad M_B = 6R_D$

Case (2): Unit load to the right of B:

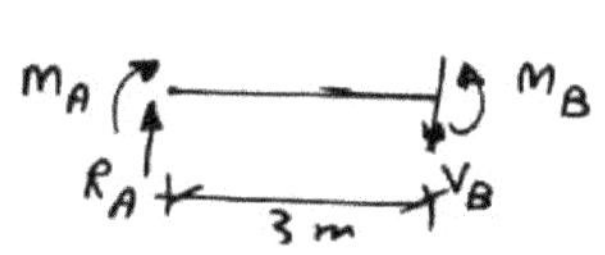

$+\uparrow \Sigma F_y = 0: \; V_B = R_A$

$+\circlearrowleft \Sigma M_B = 0: \; M_B = M_A + 3R_A$

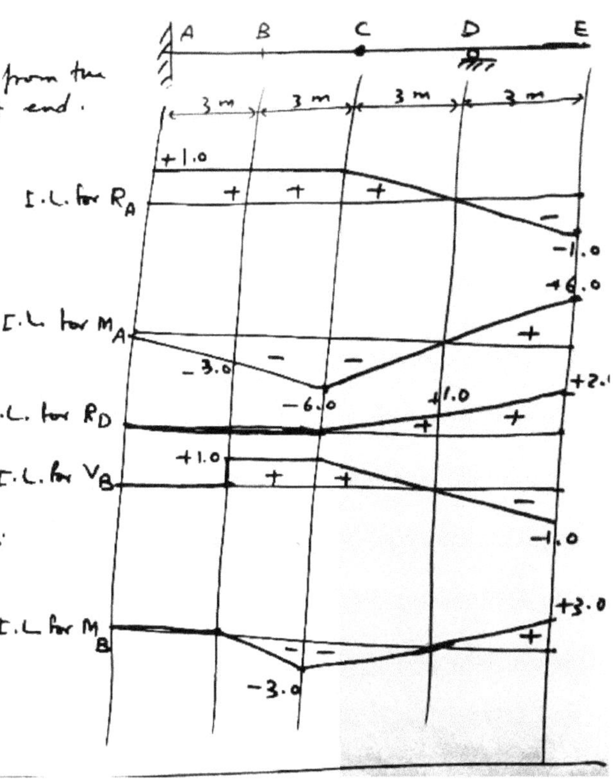

③ Consider the simply-supported beam and the wheel loads of the car shown in the figure. Determine:

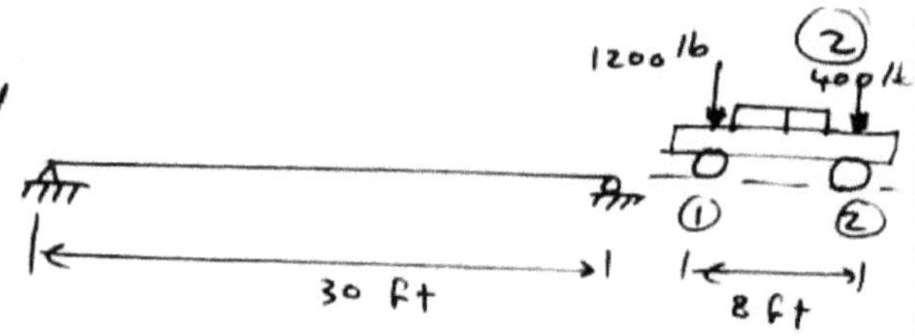

(a) the maximum moment at midspan of the beam.

(b) the absolute maximum moment and the exact position of the car on the beam.

Solution:

ⓐ Wheel ① on C:

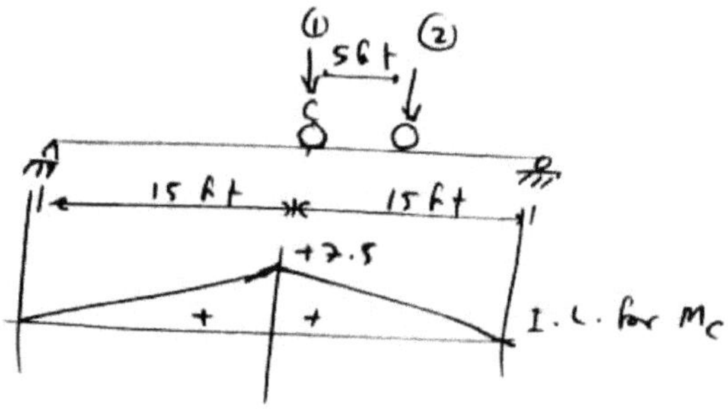

(i) just to the right of C:

$W_L = 0$, $W_R = 1600$ lb.

$\frac{W_L}{a} = \frac{0}{15} = 0 < \frac{W_R}{b} = \frac{1600}{15} = 106.7$

(ii) just to the left of C:

$W_L = 1200$, $W_R = 400$

$\frac{W_L}{a} = \frac{1200}{15} = 80 > \frac{W_R}{b} = \frac{400}{15} = 26.7$.

$M_C = \frac{1}{2}(15) = 7.5$

$1 = \frac{y}{15} + \frac{y}{15} \Rightarrow \frac{2y}{15} = 1 \rightarrow y = 7.5$

∴ wheel ① satisfies the criterion.

Wheel ② obviously does not satisfy the criterion.

∴ Max. moment at C (midspan) occurs when wheel ① is on C.

$$\therefore M_{C_{max.}} = 1200(7.5) + 400\left(\frac{7}{15}\right)(7.5) = 10400 \text{ lb-ft}$$

$$= 10.4 \text{ k-ft}.$$

ⓑ Absolute Maximum Moment:

$$\bar{x} = \frac{1200(0) + 400(8)}{1200 + 400} = 2 \text{ ft}.$$

Now, we determine the position when absolute maximum moment occurs. (It must occur under wheel ①).

$$x = \frac{L}{2} + \frac{a}{2} = \frac{30}{2} + \frac{2}{2} = 16 \text{ ft}.$$

Wheel ① at O:

(i) just to the right of O:

$\frac{W_L}{a} = \frac{0}{14} < \frac{W_R}{b} = \frac{1600}{16}$ ✓

(ii) just to the left of O:

$\frac{W_L}{a} = \frac{1200}{} = \quad > \frac{W_R}{b} = \frac{400}{16}$

wheel ① satisfies the criterion.

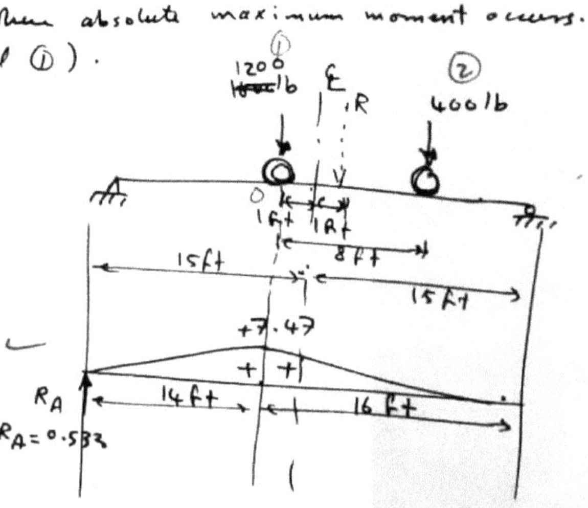

Absolute Maximum Moment

$$= 1200(7.47) + 400\left(\frac{8}{16}\right)(7.47)$$

$$= 10458 \text{ lb-ft} \approx 10.46 \text{ k-ft}.$$

(3) For the symmetric three-hinged arch loaded as shown in the figure, determine:

(a) the horizontal and vertical reactions at the supports A and C.

(b) the normal force, shear force, and bending moment at point D.

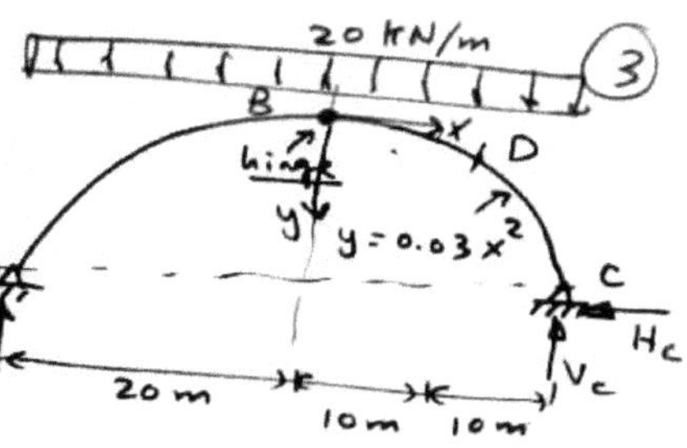

Solution:

at $x = 20$, $y = 0.03(20)^2 = 12\,m$.

From symmetry, $V_A = V_c = \frac{20(40)}{2} = 400\,kN \uparrow$.

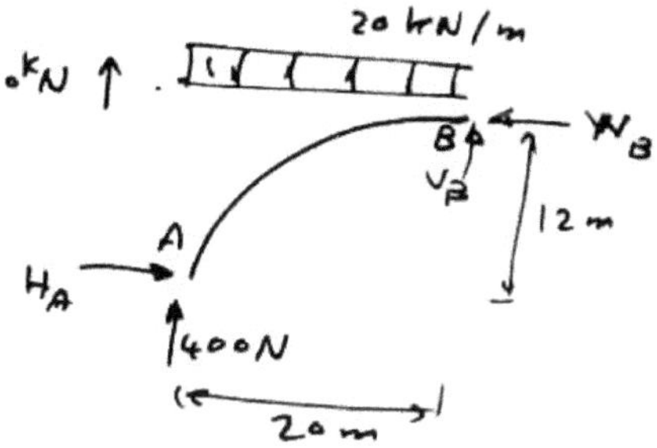

$+\circlearrowleft\ \Sigma M_B = 0:$

$20(20)(10) + H_A(12) - 400(20) = 0$

$H_A = 333.3\ kN \rightarrow$.

$\therefore\ H_c = 333.3\ kN \leftarrow$.

At point D:

$x = 10\,m$, $y = 0.03(10)^2 = 3\,m$.

Vertical Distance of D from C is $12 - 3 = 9\,m$.

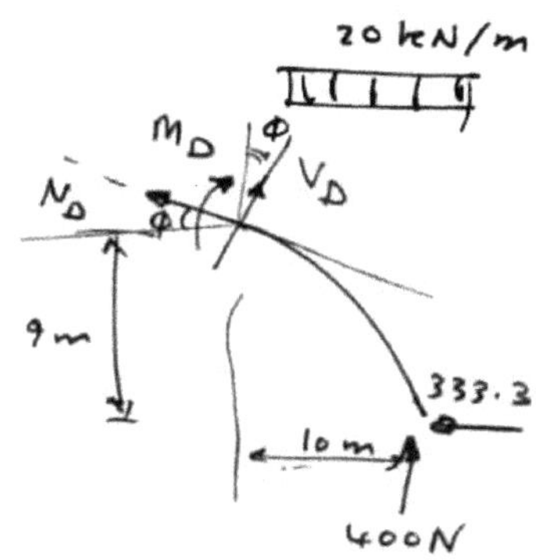

$\frac{dy}{dx} = 2(0.03)x = 0.06x$

at $x = 10\,m$, $\tan\Phi = \left.\frac{dy}{dx}\right|_{x=10} = 0.06(10) = 0.6$

$\Longrightarrow\ \Phi = 30.96^\circ \approx 31^\circ$.

$\xrightarrow{+}\ \Sigma F_x = 0:$

$-N_D\cos 31^\circ + V_D\sin 31^\circ - 333.3 = 0$ —— (1)

$+\uparrow\ \Sigma F_y = 0:$

$N_D\sin 31^\circ + V_D\cos 31^\circ + 400 - 20(10) = 0$ —— (2)

$\therefore\ -N_D\sin 31\cos 31 + V_D\sin^2 31^\circ - 333.3\sin 31^\circ = 0$

$N_D\sin 31\cos 31^\circ + V_D\cos^2 31^\circ + 200\cos 31 = 0$

$V_D - 333.3\sin 31^\circ + 200\cos 31^\circ = 0 \Longrightarrow V_D = -0.23 \approx 0 \Longrightarrow \boxed{V_D = 0}$

Substitute in (1): $N_D = \frac{-333.3}{\cos 31^\circ} = \boxed{-388.8\,kN}$ (compression).

$+\circlearrowleft\ \Sigma M_D = 0:\ -M_D + 20(10)(5) + 333.3(9) - 400(10) = 0$

$\Longrightarrow\ M_D = -0.3 \approx 0 \Longrightarrow \boxed{M_D = 0}$

Structural Analysis (1) – Exam – 3 - a – Solutions

1) Consider the compound beam shown in the figure. Draw the influence lines for:

(a) the reaction at A
(b) the shear at D.
(c) the moment at D.
(d) the shear at E.
(e) the moment at E.

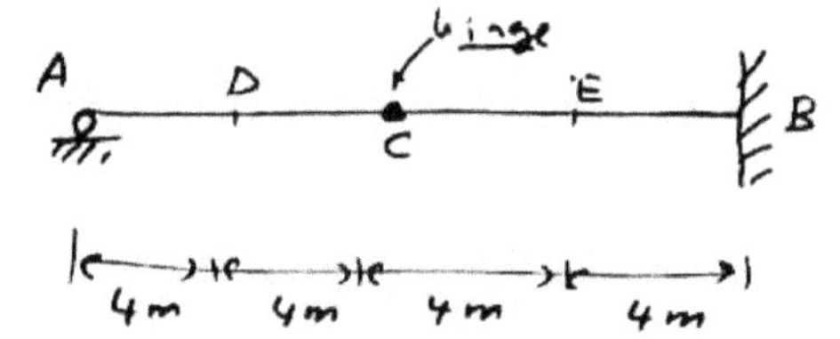

Solution:

(a) Reaction at A:

(obvious).

(b) & (c): shear and moment at D:

(i) load to the left of D:

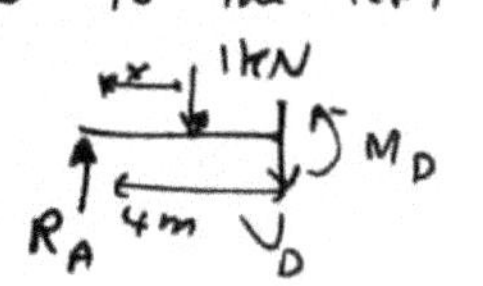

$V_D = R_A - 1$

$M_D = 4R_A - 1(4-x)$

(ii) load to the right of D:

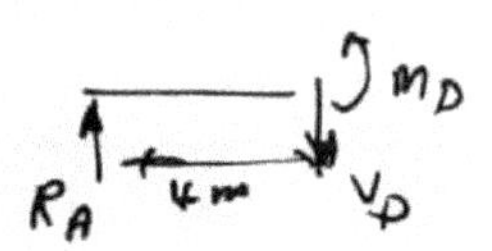

$V_D = R_A$

$M_D = 4R_A$

(d) & (e): shear and moment at E:

similar equations ~~as before~~:

(i) load to the left of E:

$V_E = R_A - 1$

$M_E = 12R_A - 1(12-x)$

(ii) load to the right of E:

$V_E = R_A$

$M_E = 4R_A$

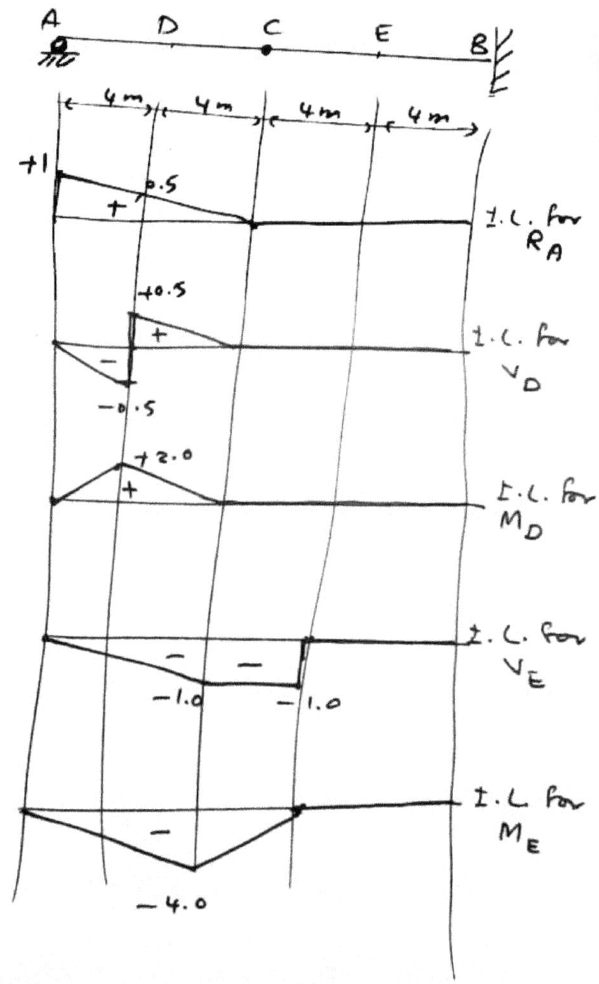

2

(2) For the simply supported beam shown in the figure, find the 'absolute maximum bending moment due to a truck, which is represented by the series of concentrated loads shown in the figure, and which moves from the right to the left.

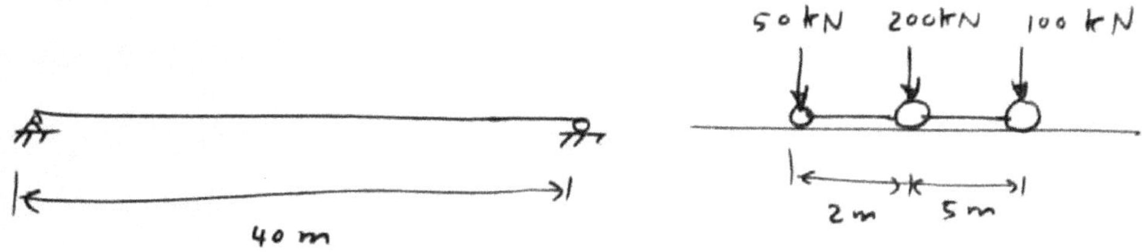

Solution :

$$\bar{x} = \frac{50(0) + 200(2) + 100(7)}{50 + 200 + 100} = 3.14\,m.$$

$\therefore a = \bar{x} - 2 = 3.14 - 2 = 1.14\,m.$

$\therefore x = \frac{L}{2} + \frac{a}{2} = \frac{40}{2} + \frac{1.14}{2} = 20.57\,m.$

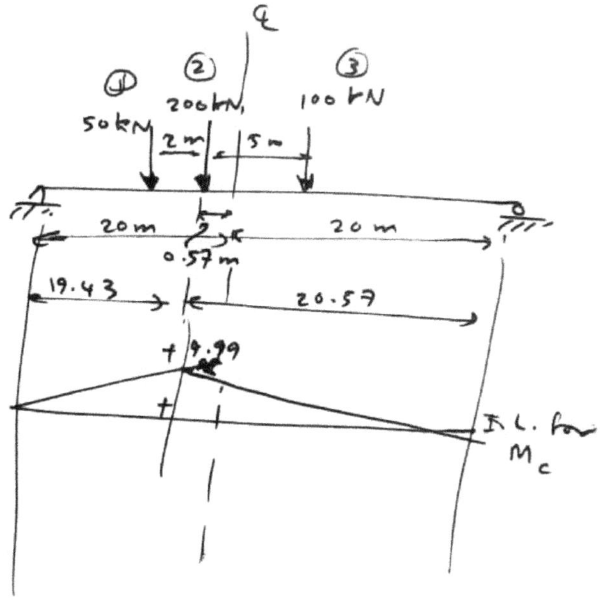

Check for wheel ② at C :

(i) just to the right of C :

$$\frac{W_L}{a} = \frac{50}{19.43} = 2.57 < \frac{W_R}{b} = \frac{300}{20.57} = 14.58$$

(ii) just to the left of C :

$$\frac{W_L}{a} = \frac{250}{19.43} = 12.87 > \frac{W_R}{b} = \frac{100}{20.57} = 4.86$$

$\therefore$ wheel ② satisfies the criterion.

$\therefore$ Absolute Maximum Moment

$$M_{abs.\,max.} = 200(9.99) + 100\left(\frac{15.57}{20.57}\right)(9.99) + 50\left(\frac{17.43}{19.43}\right)(9.99)$$

$$= \underline{3202.3}\ kN \cdot m.$$

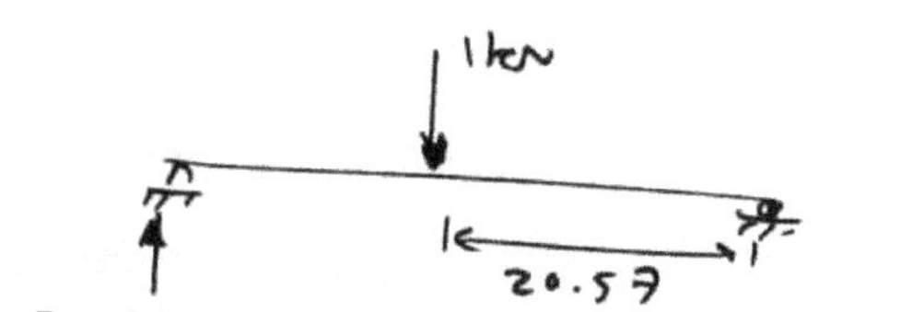

$\frac{20.57}{40}(1) = 0.514\,kN$

$\therefore M_c = 0.514(19.43) = 9.99\ kN \cdot m$

(3.) Calculate the reactions for the three-hinged arch shown in the figure. [3]
(20%).

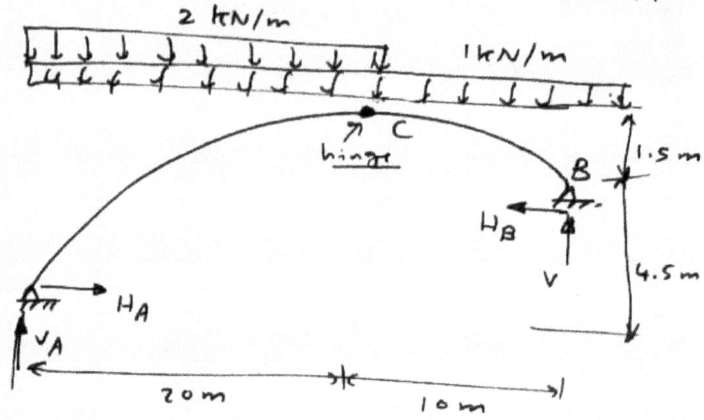

Solution:

$+\circlearrowleft\ \Sigma M_A = 0$:

$$V_B(30) + H_B(4.5) - 1(30)(15) - 1(20)(10) = 0$$

$$30V_B + 4.5H_B = 650 \quad \text{——— (1)}$$

Section at the hinge:

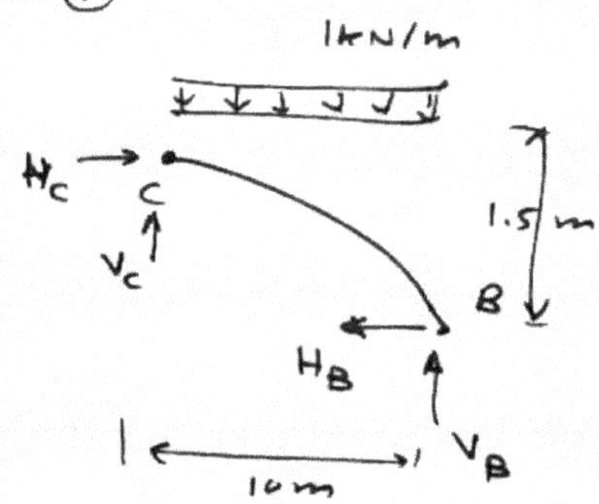

$+\circlearrowleft\ \Sigma M_C = 0$:

$$V_B(10) - H_B(1.5) - 1(10)(5) = 0$$

(3)/ $10V_B - 1.5H_B = 50$ ——— (2)

Solve equations (1) & (2) simultaneously:

$$30V_B + 4.5H_B = 650$$

$$30V_B - 4.5H_B = 150$$

$$60V_B = 800 \Rightarrow V_B = 13.33\ kN \uparrow.$$

From eq. (2): $H_B = \dfrac{10V_B - 50}{1.5} = 55.56\ kN \leftarrow$.

For the entire arch:

$\uparrow \Sigma F_y = 0$: $V_A + 13.33 - 1(30) - 1(20) = 0$

$V_A = 36.67\ kN \uparrow$.

$\xrightarrow{+} \Sigma F_x = 0$: $H_A - 55.56 = 0 \Rightarrow H_A = 55.56\ kN \rightarrow$.

www.ingramcontent.com/pod-product-compliance
Lightning Source LLC
LaVergne TN
LVHW070857210926
848482LV00016B/957